THE BEST OF FINE HOMEBUILDING

ENERGY-EFFICIENT
BUILDING

THE BEST OF FINE HOMEBUILDING

ENERGY-EFFICIENT
BUILDING

The Taunton Press

Cover photo: Nancy Hill
Back cover photos: Steve Mindel (left); Roe A. Osborn
(center and right)

Taunton
BOOKS & VIDEOS
for fellow enthusiasts

Printed in the United States of America
10 9 8 7 6 5 4 3 2 1

Fine Homebuilding® is a trademark of The Taunton Press, Inc., registered in the
U.S. Patent and Trademark Office.

The Taunton Press, Inc.
63 South Main Street
P.O. Box 5506
Newtown, Connecticut 06470-5506
e-mail: tp@taunton.com

Distributed by Publishers Group West

Library of Congress Cataloging-in-Publication Data

Energy-efficient building : the best of Fine homebuilding.
    p. cm.
    Includes index.
    ISBN 1-56158-340-5
    1. Dwellings—Energy conservation. I. Fine homebuilding.
TJ163.5.D86E52256 1999
644—dc21   99-29608
                                        CIP

Contents

Introduction

THE OTHER MORNING as I took my shower, I couldn't help but notice that none of the water was draining out of the tub. By the time I finished I was sloshing around in ankle-deep water, and my plans for the morning had changed.

Turns out I had forgotten my annual fall chore of closing the crawlspace vents, which are supposed to keep the humid summer air from accumulating under the house. This being January in New England, the vents were now funneling frigid subzero winds across the bathtub trap, which was exposed because mice had so thoroughly colonized the fiberglass insulation that it now hung from the framing like big pink stalactites. Houses are more complicated than they used to be.

When we brought the plumbing inside and started insulating our houses, they stopped being simple shelters and became systems that have to be integrated and managed. You want to keep out the cold and keep in the heat (or vice versa if you live in a cooling climate). You want to exhaust the moist air from boiling pasta, steaming showers, and tumbling clothes dryers, but you don't want negative pressure to back-draft combustion gases from the furnace into the nursery. On the other hand, you probably do want argon gases filling the space between your dual window panes. You also want housewrap, a vapor barrier, and even a self-adhering bituminous membrane up on your roof. See what I mean: complicated.

Fortunately, some of the country's best builders, architects, and engineers are willing to share what they've learned about energy-efficient building. Their thoughts and experiences are represented in this collection of articles from past issues of *Fine Homebuilding*.

—*Kevin Ireton, editor*

Testing Homes for Air Leaks

A blower-door test is a simple, economical method of learning how tight a house is and where the leaks are

by Marc Rosenbaum

To reduce the amount of energy used in heating and cooling their homes, most people focus on adding insulation. But controlling air infiltration is more important. Airtight homes cost less to heat and cool and are more comfortable to occupy (as long as the homes are equipped with mechanical ventilation systems). Airtight homes are more durable, too. In leaky houses, moist, heated air escapes to cooler areas, such as insulated walls and ceilings, and the vapor condenses, causing mold, mildew, rot and other moisture-related problems.

Still, few builders know how tight the homes they build are or where the major leakage occurs. A blower-door test (photo right) can provide a builder with this information. In this article I'll explain what's going on when an energy consultant or an insulation contractor does a blower-door test. I'll also explain how to do the test in case you're a builder who specializes in energy-efficient construction and are considering buying a blower door.

Enter the blower door—Houses leak air because of the difference between indoor and outdoor air pressures. A blower door places a home under a known, elevated pressure and then measures how much airflow is required to maintain the pressure difference between indoors and outdoors. The tighter the house, the less air the blower door must move to maintain a given pressure.

Blower doors were first developed in Canada and Sweden as research tools, but their use by builders, insulation contractors and weatherization crews is gradually increasing. In new construction, for example, a blower door is a quality-assurance tool, enabling the crew to measure the effects of their air-sealing efforts and to see where they've succeeded and where they have further work to do. In existing buildings the blower-door testing process can determine the potential for savings through weatherization and

Blower-door test. **A blower door is sealed within an exterior-door opening. A fan at the bottom of the blower-door assembly pulls air from the house, and the digital manometer clamped on the door measures the difference between indoor and outdoor air pressures. A blower-door test tells how tight a house is, and the blower door also helps pinpoint air leaks.**

can direct the weatherization crew to the leaky areas.

Components of a blower door— Portable, rugged and affordable blower doors have been around for more than a decade. Today, a typical blower door consists of a fan, a frame-and-panel assembly that fits into an exterior-door opening and some instrumentation. My Minneapolis Blower Door from the Energy Conservatory (see sidebar p. 10) includes a 20-in. dia., variable-speed, reversible ¾-hp fan, an aluminum frame (top photo, right) and a fabric panel. I drape the panel over the frame and adjust its lever-operated cams so that the assembly fits tightly in an exterior-door opening. Weatherstripping around the blower-door frame makes a tight seal. Then I slip the fan housing into an elastic-rimmed opening in the blower-door panel.

Once the blower door is installed, I mount the fan-speed controller and the instrumentation that measures air-pressure levels (I use a digital manometer) nearby, often clamped to the open door. A tube runs from the manometer to the fan housing, and another runs through the blower-door panel to the outside. The manometer enables me to measure the pressure difference between inside and outside, as well as the pressure across the fan, which correlates with airflow rate.

Houses vary tremendously in how tight they are, so the amount of air that a blower-door fan has to move to maintain a given pressure difference also varies. Therefore, fan housings are usually designed so that you can reduce their flow area (the size of the fan opening). Retrotec (see sidebar p. 10) makes a door that has a low-flow plate with a number of holes that can be plugged for testing tighter homes. My Minneapolis door has three concentric low-flow rings that snap into the fan housing, giving a choice of four ranges in all. The rings allow accurate measurement of flow rates ranging from 100 cu. ft. per minute (CFM) to 6,000 CFM, a typical range for blower doors.

Performing a test— Before turning on the fan, I quickly prep the house for a blower-door test. I close all exterior windows and doors, but I leave interior doors open (doors to the basement may be left open or closed, depending on the goals of the test). I turn off combustion appliances, exhaust fans, dryers, air conditioners, etc. I close fireplace dampers and doors. I may tape off certain intentional openings, such as the exterior

Frame within a frame. **Although it comes with a wooden frame, one of the optional components of a Minneapolis Blower Door is this adjustable aluminum frame. Lined with weatherstripping, the frame has cam levers that allow you to fit it tightly into an exterior-door opening.**

Finding the leaks. **After reversing the fan to pressurize the house, the author tests for leaks with this Teflon squeeze bottle, which contains a chemical that combines with air to produce a puff of smoke. Here, smoke escapes beneath an exterior-door saddle. If the door saddle had been caulked correctly, the smoke would just hang in the air.**

hoods of a heat-recovery ventilator, so that the house doesn't appear leakier than it really is.

If a breeze or a strong temperature difference between inside and outside is creating a difference between indoor and outdoor air pressures, I adjust the pressure gauge to read zero. This adjustment takes the weather conditions into account and allows me to measure accurately the house pressure.

Many weatherization crews conduct a quick, single-point test at 50 pascals (a pascal is a metric unit of pressure—6,895 pascals is 1 psi). The test tells them whether the house is leaky enough to justify the cost of air-sealing it (making a home super tight).

A more-accurate test, and the one I do, is a multipoint depressurization test. I bring the fan speed

up slowly, depressurizing the house to 60 pascals below the outdoor pressure. I note the fan pressure in pascals. Then I slow the fan, bringing the house pressure to 50 pascals, and record the fan pressure. I continue to repressurize the house to 40, 30 and 25 pascals, noting the fan pressure at each point.

When I am through (it takes about three minutes to do this test), I enter the data, along with the indoor and outdoor temperatures, the volume of the house, its exterior surface area and the number of stories, into the blower-door program on my portable computer. The program also asks whether the house is sheltered, exposed or average in terms of its exposure to the wind.

The computer then displays the natural air-change rate for the house based on three different measurements: First, it calculates the CFM50, which is how many cubic feet per minute of air the blower door moves to maintain a 50-pascal differential between indoors and outdoors. Older homes typically range from 2,000 to 4,000 CFM50, although some exceed the 6,000 CFM50 capacity of the blower door, in which case I use a can't-reach-50 factor to estimate the CFM50. Values for new homes are lower, usually 750 CFM50 to 2,500 CFM50, due to better windows and doors, platform framing and the use of sheathing panels instead of boards. Finally, homes that have been air-sealed frequently achieve CFM50 figures of 250 to 750. The tightest homes test even lower.

The second calculation the computer shows is the ACH50 (air changes per hour at 50 pascals), a value indicating how many times in one hour the volume of air in the house can be pumped through the blower door at 50 pascals. The ACH50 value is one way to compare houses of different sizes. Older homes commonly have ACH50 values of 10 to 15; newer homes typically experience about 4 to 10 ACH50, and houses that have been air-sealed experience 1 to 3 ACH50. The Swedish standard for new single-family homes is 3 ACH50. The Canadian R-2000 program, a program for energy-efficient new residential construction, has a maximum leakage standard of 1.5 ACH50.

The third calculation shown on the computer is the effective leakage area (ELA), which combines all of the air-leak holes in the house into one single opening. The ELA helps the user envision the size of the hole to be sealed. My own home has an ELA of just over 12 sq. in. A home I tested last winter had an ELA of 473 sq. in., or the size of a small, open window.

Estimating natural infiltration— The blower-door test is conducted at pressure differences that exceed significantly the pressures a house experiences under everyday conditions. One method that's widely used to convert blower-door results to estimate natural infiltration was developed at the Lawrence Berkeley Laboratory. This method takes the ACH50 and divides it by a correlation factor that considers climate (average wind speed and outdoor temperature), the building's height, its protection from the wind and a leakiness correction factor that's based on an estimate of the type of holes predominant in the home—small cracks (tight home) or large holes (loose home). In most cases this correlation factor falls between 14 and 20, so a home whose ACH50 is 10 would have a predicted natural infiltration rate of 0.71 ACH to 0.50 ACH.

This correlation is only an estimate, however; recent work in the northwestern United States compared the correlated ACH50 calculations to tracer gas measurements of natural air leakage (one of the most accurate methods) and concluded that the correlation has a +/-50% margin of error. I rely on the CFM50 figures calculated from the blower-door test, which are good to +/-5%, and give me a better idea of how one house compares to another, regardless of its size.

Sealing leaky areas—The contractor should have various air-sealing tools on hand when the blower-door technician shows up so that many of the problem areas can be fixed right then. Tools for air-sealing include a foam gun, acrylic adhesive tape commonly used to tape air barriers, acoustical sealant, polyurethane caulk, polyethylene sheet and some scraps of rigid foam. The foam gun, the tape, the poly and the sealants are all available from the Energy Federation, Inc. (14 Tech Circle, Natick, Mass. 01760-1086; 800-876-0660.)

Any exhaust fan, including dryers and range hoods, could draw some of its makeup air down the chimney of a boiler, furnace or water heater, a phenomenon called backdrafting. So when a weatherization crew conducts a blower-door test, the crew should also do a combustion-safety test to make certain the house can be safely air-sealed without increasing the risk of backdrafting into the house.

Picking the best time to test in new construction is a trade-off between having the wiring, plumbing and ventilation penetrations already in place and being able to get at problem areas to fix them. I think the right time to test is after the drywall is taped and before interior trim has been installed. At this point, leaks around rough openings can be sealed, and any holes between the living space and the attic are still accessible.

To find the leaks, I use a smoke generator (bottom photo, p. 9). It's a small, Teflon squeeze bottle containing mineral wool and titanium tetrachloride (TiCl4). The TiCl4 combines with moisture in the air to make smoke that's the same density as air, so the smoke doesn't rise. I'm careful not to breathe it—the smoke is corrosive and not recommended for lung maintenance. I squirt a small puff of smoke toward a suspected leak. If I'm in a space pressurized by the blower door, the smoke will find any leakage path where air is headed to lower pressure.

Because the biggest pressure differences are high and low in a building, holes in these areas typically leak more air than holes at the middle do. So I concentrate a leak search at the attic and the basement.

I seal high holes first. If the house has an unfinished attic, I depressurize the house (it's easier to see smoke going out than blowing in) to 20 to 25 pascals, go into the attic and blow smoke at all the potential air-leak sites. I pay particular attention to gaps around chimneys, duct chases, holes for plumbing stacks and wiring, attic hatches and recessed fans, lights and ceiling junction boxes. Other big leakers include dropped ceilings, kitchen soffits and kneewalls in one-and-a-half story homes.

After plugging the attic holes, I reverse the blower-door fan to pressurize the house and move on to the basement or crawlspace. Common leakage areas are the sill/foundation joint, the rim joist/sill joint and utility penetrations.

Finally, I find holes within the living area. I aim the smoke at potential leakage sites, such as the top and bottom plates of the walls (plates tend to leak), around windows and doors (bottom photo, p. 9) and electrical penetrations. It's not uncommon to see smoke zipping into an electrical outlet on an interior partition wall. The wires may go up to the attic or to an exterior wall and then to an exterior outlet, or the leakage path may be more complex.

Other uses—As an energy consultant, I bring my blower door with me on troubleshooting calls. Using the blower door on houses with moisture problems, I can determine whether the problem is excessive moisture sources or inadequate ventilation. Assessing air leaks between the house and the attic, for instance, can help locate the source of an ice dam, which often is produced by warm house air leaking into the attic and warming up the roof sheathing.

Blower doors can diagnose other building problems. Increased awareness of problems caused by leaks in air-distribution ducts, such as wasted fuel, indoor air pollution and damage caused by pressure-driven moisture, has fostered a number of ways to assess duct leakage with the blower door. In the simplest method, the air leakage of the house is measured twice, once with grills and registers open and once with them sealed. This method is relatively quick but probably the least accurate. Minneapolis Blower Door has recently come out with the Duct Blaster, a small fan that assesses duct leakage directly.

Recently, weatherization specialists Michael Blasnick and Jim Fitzgerald of Philadelphia, Pennsylvania, and Minneapolis, Minnesota, respectively, have used the blower door and the space-to-space pressure-differential measurements to assess how leakage occurs between heated space, buffer spaces such as attics, kneewalls and basements, and the outdoors. Their methods help determine where significant leakage is occurring in complex leakage paths. □

Marc Rosenbaum, P. E., owns Energysmiths, which designs low-energy use, environmentally sound housing. Photos by Rich Ziegner.

The Minneapolis Blower Door from The Energy Conservatory.

Shopping for a blower door

Four companies make blower doors in North America. When shopping for a blower door, look for these features:

Ease of portability and setup—The first doors were heavy and cumbersome. Some products available today break down into convenient pieces (photo above) and assemble quickly.

Range of airflow and pressure—If you plan to test only airtight new construction, you might get away with a door like the Duct Blaster, which only goes up to about 1,500 CFM. If your work includes existing homes, you'll need a unit with 6,000-CFM capability.

Accuracy and repeatability—The inaccuracies of a blower door are caused by the manufacturing tolerances of the fan and the capabilities of the instrumentation. A typical setup will be accurate to +/-5%. With better instrumentation (I bought a digital manometer for my door to replace the standard gauges), this accuracy may improve to +/-3%. The repeatability accuracy—testing a home a bunch of times and getting the same results—of a blower door should be +/-1% to 2% in low-wind conditions.

Features and price—Check which features are standard and which are extras, such as knock-down frames, hand-held computers, software, sophisticated instrumentation and/or carrying cases. Prices start at about $1,350, and a fully equipped blower door goes for about $2,000.

Sources of supply
The Energy Conservatory
5158 Bloomington Ave. So.
Minneapolis, Minn. 55417
(612) 827-1117, fax (612) 827-1051.

Infiltec
P. O. Box 8007,
Falls Church, Va. 22041
(703) 820-7696.

Retrotec Energy Innovations, Ltd.
1504 Merivale Road
Ottawa, Ont., Canada K2E 6Z5
(613) 723-2453, fax (613) 723-2574.

Eder Energy
7535 Halstead Dr.
Mound, Minn. 55364
(612) 446-1559.   —M. R.

Understanding Common Moisture Problems

Three case studies illustrate the importance of controlling humidity and air leakage

by Marc Rosenbaum

About a dozen years ago, a couple called me to complain about serious water stains on the kitchen ceiling of their new home. The builder and the architect were at each other's throats: The builder blamed the stains on the polyethylene vapor retarder the architect had insisted he install in the ceiling, and the architect disagreed but had no alternative explanation. The confused homeowners hoped that I could offer some help.

It took less than a minute to identify the source of the water causing the stains. Six recessed lights punched holes in the ceiling, and they acted like little chimneys, transferring moist kitchen air into the attic, where the vapor condensed on the cold roof sheathing and dripped back down to the ceiling drywall.

Once everybody could see the evidence, they understood what was occurring and could agree that the solution was to seal the recessed lights.

What I learned that day was how much myth and dogma exist in the design and construction professions about simple, common building failures that have straightforward physical explanations.

In the three case studies that follow, I describe the nature of the problems encountered, the diagnostic methods and tools used to determine the causes, and the recommended fixes. All of these homes are in the Northeast, but the construction practices that caused the failures are common throughout the country. What varies is the type of problem that results. In the end, most residential failures are caused by uncontrolled movement of air and/or moisture, whether the building is in Mississippi or Minnesota.

First, find out exactly what's wrong— Occasionally, I may be able to diagnose a straightforward problem over the phone. But if I visit the house, first I get a thorough description of the phenomena. I want to understand what is happening, where in the house it occurs, in which seasons and how long it has been going on. This last point is important; many problems are the result of a chain reaction that follows a change in a building, like a new kitchen, furnace or windows. Any of these things might alter the humidity level or create new pathways for air.

The next step is a thorough walk-through. The homeowner may have called me because of something obvious like severe condensation on windows. But a troubleshooter might find other failures, indicating a wider problem. So my rule is to start the examination at the footing and end at the ridge.

Case #1: Frost, water stains and ice dams— In the first of the three case studies, the builder of a 3-year-old house called me because he had been unable to solve wintertime problems of severe frost buildup in the attic, stains on the second-floor ceiling below and recurrent ice dams.

The house had a gas-fired, forced-air heating system with central air conditioning and was located in a 6,500-heating-degree-day climate. (Degree days, a measure of heating demand, are calculated by subtracting the average daily outdoor temperature from a designated base temperature, typically 65°F. A day with an average

The low outdoor temperature meets high indoor moisture. Condensation on this window could be prevented by lowering the relative humidity inside the house, increasing the window's R-value or both.

Excessive indoor humidity causes mildew growth and peeling paint. Water running off this window has saturated the sash, damaging its finish. The moisture content of wooden windows in this passive-solar house was as high as 28%, as compared with the normal level of around 10%.

temperature of 40°F, for example, would be a 25-heating-degree day. Annual figures are simply the sum of the daily figures. For comparison, San Francisco averages about 3,000 degree days and Chicago 6,500.)

The insulation was kraft-faced fiberglass batts, exceeding code-required levels, and no special air-sealing measures were implemented during construction. Soffit vents and a ridge vent, properly installed, provided adequate roof ventilation. The contemporary design yielded two separate attic spaces above second-floor bedrooms, separated by a loft, and another attic above the garage. A walk-through showed a dry basement and no evidence of water in the house except for the minor ceiling stains.

Keep heat in the living space—Ice dams are caused by warming the underside of the roof, causing snow above to melt, run down the roof and freeze again where the roof temperature drops below freezing, commonly at the eaves. (For more on ice dams, see *FHB* #98, pp. 60-63.) Water backs up behind the ice dam and leaks into the building. Although the classic solution for ice dams is to add insulation and/or roof ventilation, I look first for sources of unintentional attic heat that is warming up the roof sheathing. In this house, I didn't have far to look.

On a day that was substantially below freezing, I measured temperatures of 47°F in the attic above one second-floor bedroom and 40°F above the other. Because an insulated, ventilated attic should be only a few degrees above the outdoor temperature, we were in prime ice-dam territory. We began to look for sources of warm air leaking to the attic or locations of inadequate insulation. We found both.

A leaky building envelope—The kneewall access panels from the loft to the two attics were poorly sealed and crudely insulated with fiberglass batts duct-taped to their backsides. The door to the attic above the garage was uninsulated but had weatherstripping at the latch side and top. The fiberglass insulation in the walls between the heated spaces and these attics was open on the backside to the attic. This practice is common, but the effectiveness of fiberglass insulation is compromised when it is installed in an open cavity: Warm air adjacent to the backside of the drywall rises and moves into the attic, being replaced by cold attic air at the bottom. Fiberglass by itself does not stop air movement. Indeed, one clue a troubleshooter should always watch for is dirty fiberglass insulation. Fiberglass is a great filter, and dirty fiberglass has had a substantial amount of air moving through it.

In the case at hand, the major source of heat in the attic came from a different source. The heating ducts for the second-floor rooms were in the attic, and they were unsealed. The metal ductwork was not insulated, and the flexible ductwork had 1 in. of fiberglass insulation, not much for ducts carrying 120°F to 130°F air in what is nominally an outdoor space (center photo). In addition, the attic-duct trunk came up through a basement chase that was unsealed top and bottom. This chase allowed warm basement air (the uninsulated ductwork kept the basement toasty) to rise freely into the attic. But what about the frost on the attic roof sheathing? Many homes have ice dams without this symptom also appearing. Frost occurs when water vapor in the air hits a cold surface and condenses or freezes on that surface. To solve this problem, we needed to find how moisture was entering the attic.

High-tech tools aid the search—With my sling psychrometer (sidebar p. 15), I measured the relative humidity in the house: It was 50%, higher than the 35% to 40% I prefer to see during the winter. A quick blower-door test (see *FHB* #86, pp. 51-53) showed an airflow rate of 2,375 cu. ft. per minute (CFM) at a 50-pascal pressure difference (CFM50). (A pascal is a metric unit of pressure; 6,895 pascals is 1 psi.) This amount is fairly typical of new construction with no special attempt to air-seal. The house was in fact probably a bit better than the average new home with forced-air heat. These houses tend to be leakier than homes with hot-water heat because of leaky ductwork. With normal amounts of moisture be-

Paint doesn't adhere well to a moving surface. Wood siding that has not been back-primed or otherwise sealed can expand and contract as it takes on and gives off moisture. As the wood siding moves, its painted finish cracks and peels.

Losing heat on the way to the living space. Heat-supply ducts in this attic were poorly sealed and underinsulated. As a result, the temperature inside the attic was 47°F, substantially warmer than the outside air. This situation prompted the formation of ice dams.

The source of moisture wasn't hard to find. This clothes-dryer duct, torn and hanging by a wire rib, is supplying warm, moist air directly to the attic.

ing generated in the house, this level of leakiness would typically result in a lower relative humidity, so I suspected an unusual moisture source somewhere. Bath fans, dryer and range hood were vented outdoors. Once again, the heating system was the major culprit.

Mounted on the furnace plenum was a central humidifier. Although it was on a low setting, the builder reported that it had only recently been turned down. I suspected that this home had been running at relative-humidity levels exceeding 60%, which is unhealthy for both the building structure and the occupants (many people are allergic to dust mites, which need high humidity to flourish). The central humidifier added moisture

to the heating air, some of which leaked directly into the attic. A more effective method for moisturizing the attic could hardly be devised.

The proposed solution is twofold—I recommended that all attic ductwork first be sealed with latex mastic and then well-insulated. This task could be accomplished by covering the ductwork with loose-fill cellulose, a type of insulation made from recycled newspapers that have been shredded and treated with a fire retardant. Another alternative would have been to create a sealed duct chase out of foil-faced fiberglass duct board, laying the flexible duct in it and backfilling the space around the ducts with cellulose. I

advised the homeowners that the central humidifier be disconnected.

Recommended fixes to the building itself included sealing the duct chase at both basement and attic levels with a rigid material such as plywood or sheet metal, sealing the house-to-attic penetrations (plumbing stacks, radon stack, electrical), installing an air barrier on the backside of the kneewalls (drywall or housewrap), and weatherstripping and insulating the door and kneewall hatches to the attics. In addition, I suggested installing a fan in the master bath, vented outdoors, and sealing the basement ducts with latex mastic. These changes not only will solve the problems and increase general durability of the home, but also will reduce the home's energy use significantly.

Case #2: A passive-solar house with severe moisture problems—Our second case study is a 10-year-old, passive-solar house with a south-facing, finished walk-out basement in a 6,000-heating-degree-day climate. The problems were severe condensation and frost on the windows and glass doors, causing mildew and paint peeling (top photo, p. 11); mold in the lower corner of one of the basement bedrooms; water stains around recessed ceiling lights; stuffy air quality; and peeling exterior paint. The home had an electric furnace, forced-air heating system and three heat-recovery ventilators, one serving the main body of the house and two small ones, each serving only a bathroom.

A walk-through showed condensation also appearing against the band joist in the unfinished basement on the house's north side. Because it was May, measuring relative humidity wouldn't necessarily have given an accurate indication of winter conditions. However, I used my moisture meter (sidebar p. 15) to assess moisture levels both inside, checking the millwork and the trim, and outside, testing the exterior siding and trim. Some of the windows had spot moisture contents as high as 28%, clearly caused by condensation (bottom photo, p. 11). Interior trim was well below the danger zone of 20% or more, where wood becomes susceptible to decay organisms. Some exterior readings also were in the 20s, mostly in the areas where the trim was too close to the ground and splash-back was occurring.

The peeling of the exterior paint appeared to be caused by water penetrating from the outside (top photo, facing page). The house had minimal overhangs on the gable ends, and the grading around the foundation put the wood too close to the ground in many places. Neither the trim nor the siding had been back-primed before installation, so water drawn up between the clapboards by capillary action could soak readily into the wood. The expansion and contraction of clapboards as they get wet and then dry eventually causes the paint film to fail. One clue suggesting an exterior problem—unrelated to high indoor-humidity levels—was that the unheated garage showed the same peeling-paint symptoms.

Ventilation wasn't as it seemed—I did a quick blower-door test and found that the house was quite tight (1,000 CFM50). Then I checked the

Insulation is no good if it's not continuous. Cold air from this crawlspace has a clear path to the floor of the adjacent bedroom through this and other uninsulated joist bays. Dirt in the insulation could mean that air is escaping from the living space.

Ceiling fixture provides light below and heat above. This recessed ceiling light, installed during a remodeling job, provides a conduit for warm, humid air to flow from the living space into the attic. The preferred alternative in cases such as this one is to install the type of light fixture that can be covered with insulation completely, without causing a fire hazard.

operation of the house's three ventilators to verify that air was in fact being exhausted. The central unit was moving hardly any air at the exterior vent hood, and one of the small units was incapable of opening the flap on its vent hood at all. The third small unit worked as it was supposed to. It was clear that the house, despite appearing well-ventilated, was suffering from too little air exchange. Humid air leaking from the house into the attic condensed on the roof sheathing and dripped back to the ceiling, causing stains.

In the unfinished basement, I taped 2-ft. squares of clear polyethylene to the concrete wall and to the basement slab. If there is a significant source of ground moisture, condensation often will bead on the backside of the poly. In this case, none was observed. Moisture-meter readings on the concrete appeared reasonable. However, I had noticed that the gutter downspouts ran into standpipes leading down to the footing drains, not a good practice for keeping basements dry. I inspected the footing drains where they ran out to daylight, and no water was running, unusual for late spring in the northeast.

The mold in the lower corner of one of the bedrooms appeared where the concrete sidewall met the wood-framed south wall. The basement's concrete walls continued beyond the building to form retaining walls on the east and west, and I strongly suspected a direct thermal bridge through the concrete. This bridge would keep the wall near the corner cold in the winter, causing a rise in the relative humidity at the wall surface. This increased relative humidity causes the moisture content of the wall surface to rise to the level where it will support growth of mold.

Moisture is the culprit—Virtually all of the interior problems could be corrected by substantially lowering winter moisture levels. This house needed reliable ventilation, so I suggested a new heat-recovery ventilator having adequate capacity (at least 100 CFM), and operating it enough to maintain 35% to 40% relative humidity in the heating season. The original ventilator was neither well-designed nor well-installed. Mold needed to be removed and the millwork repainted.

Even if the moisture is controlled, the southeast basement corner may still foster mold. This wall may need to be opened and insulated from the inside to maintain a higher interior surface temperature, and thereby a lower moisture content, not conducive to mold growth.

On the exterior, regrading would move water away from the building, and rotten trim could be replaced with wood primed on all sides. The gutter downspouts should be disconnected from the footing-drain system, and the drains should be checked for blockage. The clapboard laps could be wedged open with plastic wedges (Shur-Line Inc., 2000 Commerce Parkway, Lancaster, N. Y. 14086; 800-828-7848) and allowed to dry. (Some cracking of the clapboards is likely to occur as the wood dries out.) Once dry, they could be repainted with a good latex paint.

Case #3: Problems common to conventional construction—In the third case study, the owner of a 32-year-old home complained of room-to-room temperature variation in the winter, high utility bills and condensation in the attic. Located in a 5,000-degree-day heating climate, the house had forced-air gas heat. The second floor of this French Eclectic-style house was contained in flat-roofed dormers poking up through the steep hip roof. Kneewall spaces adjacent to the second-floor bedrooms were designed to be cold, and they were vented outdoors.

The bedrooms usually were cold—no mystery once I got in the crawlspaces. The ductwork to the bedrooms ran through the cold spaces, was leaky and had minimal insulation (bottom photo). The 2x4 kneewalls themselves were insulated with fiberglass batts, the backs of which were open to the cold. But the biggest offender was the fact that the 2x12 second-floor joist cavity was open to the kneewalls (top photo, p. 13). Cold outdoor air could flow into the kneewall cavity, through the joists and to the other kneewall cavity. The net effect was like having an uninsulated floor in contact with the outdoors.

Heat loss through the ducts—The blower-door test gave a result of over 4,000 CFM50—very leaky. I started the furnace and used a smoke pencil—a small squeeze bottle containing chemical vapors that produce a smokelike gas on contact with air—to look for duct leaks. There were plenty. A good portion of the hot air generated by

Don't forget to replace the insulation. In his work, the author commonly finds conditions such as this uninsulated bathroom exhaust fan. The insulation, cleared away to make room for the retrofit fan and duct, was never replaced.

Close the gaps. Poorly insulated heat-supply ducts raise the temperature of attics, increasing the chances of condensation and ice damming.

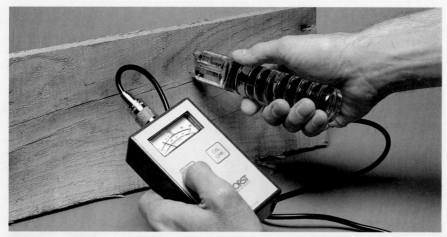

Moisture meter.

Thermometer.   **Smoke pencil.**

Sling psychrometer.   **Blower door.**

Tools for troubleshooting

I carry several instruments with me on all of my residential troubleshooting adventures. The first is a digital thermometer, useful for measuring attic temperatures, duct temperatures and the like. It costs about $20.

For measuring relative humidity, I use a sling psychrometer. This instrument consists of two thermometers, one of which has a dampened wick on the bulb. The tool is whirled around so that the thermometer bulbs are in moving air. The thermometer with the wick relates to how much moisture is in the air. The relative humidity is calculated from the two measurements. Mine is made by Bacharach Inc. (625 Alpha Drive, Pittsburgh, Penn. 15238; 412-963-2000) and costs about $70.

A blower door—an instrumented, portable fan installed in an exterior door—is the tool for assessing how tight a house is and where the leaks are. My kit includes a two-channel, digital micromanometer, which measures pressure differences between indoors and outdoors, or between rooms or floors in a home. The Minneapolis Blower Door, including the micromanometer, costs about $2,000. It's made by The Energy Conservatory (5158 Bloomington Ave. S., Minneapolis, Minn. 55417; 612-827-1117).

A smoke pencil is a small plastic squeeze bottle containing chemicals that produce a visible gas on contact with air.

Finally, a moisture meter, which measures the moisture content of wood and other materials as a percentage of dry weight, is invaluable. Mine is made by Delmhorst (Delmhorst Instrument Co., 51 Indian Lane East, Towaco, N. J. 07082; 800-222-0638). It cost about $300.—*M. R.*

the furnace was not getting into the living spaces. There was a lot of leakage in the basement ductwork and in the return side of the system, which used joist and wall cavities, some panned with sheet metal, as ducts. Another reason for the high utility bills was that the original furnace had been replaced in the late 1980s with a new unit that had a 67% efficiency rating—unimpressive at a time when furnaces were available with efficiency ratings as high as 95%.

In the attic, I found the source of moisture causing the condensation. A recent kitchen and bath remodel had added a number of recessed lights and two bathroom-ceiling exhaust fans. The insulation around each fixture and fan had been removed, and no attempt had been made to seal the ceiling penetrations (bottom photo, p. 13; top photo, facing page). Most of the insulation in the attic was black with dirt, indicating rampant air leakage from the rooms below. Both fans had been vented with 3-in. flexible plastic duct, which had been flattened so that very little air actually made it outdoors. To top things off, the dryer had been vented through the roof, also with plastic duct. The duct was torn and hanging only by a wire rib to the roof vent cover (bottom photo, p. 12). And the vent cover was screened and plugged with lint. The net result: All of the moisture from the dryer was venting to the attic.

Stem the flow of air on all fronts—To reduce air leakage in the house, I recommended the bath fans and recessed light fixtures be sealed. The insulation could be improved by filling the second-floor joists with blown-in, dense-pack cellulose insulation, or by sealing the joists from the kneewall area with rigid-foam insulation, foamed in place. The back of the kneewalls should be sealed with housewrap or drywall. Plastic sheeting should not be used on the cold or attic side of insulation because it can trap moisture inside the wall (see *FHB* #88, pp. 48-53). Another alternative would be to insulate the sloped portions of the roof with dense-pack cellulose, bringing the kneewall areas within the heated envelope.

Ductwork should be sealed with latex mastic. Return ducts using standard joist bays should be lined with metal and sealed. Insulation should be added to ductwork in kneewall areas after it is sealed. Insulating basement ductwork would increase comfort by maintaining air temperature on the long duct runs. The dryer vent should be replaced with sealed metal ductwork. The long-term strategy would include replacing the low-efficiency furnace.   □

Marc Rosenbaum, P. E., owns Energysmiths in Meriden, New Hampshire, and designs energy-efficient and environmentally sound housing. Photos by the author except where noted.

Further reading

Two excellent resources of related information are *Advanced Air Sealing* (Iris Communications, 258 E. 10th Ave., Suite E, Eugene, Ore. 97401; 800-346-0104) and *Moisture Control Handbook* by Joseph Lstiburek and John Carmody (Building Science Corp., 273 Russett Road, Chestnut Hill, Mass. 02167; 617-323-6552).

Unintentional "chimneys" let warm air into the attic. The chase containing this plumbing vent pipe extends all the way through the house and should be blocked off to prevent the loss of heated air from the attic.

Don't seal ducts with duct tape. A fiber-reinforced mastic is a better choice than duct tape for sealing the joints in ductwork because it won't pull away and fall off.

Fixing a Cold, Drafty House

Forget about weatherstripping doors and windows. Sealing and insulating the attic are the keys to lower heating bills and a more comfortable house.

by Fred Lugano

As a weatherization contractor, I meet a lot of people who are sick and tired of cold, drafty houses. Their problem—and maybe yours, too—is that they live in homes that don't work very well. These houses, new and old, cost too much to heat and to cool. Their paint peels, their roofs dam with ice, and they sometimes make their owners sick. Simply put, these homes lack a good thermal envelope, or an insulated, air-resistant boundary between conditioned inside air and outside air.

Incomplete thermal envelopes are common in old houses, but new ones can have the same or similar problems. Open-web joist systems, cantilevers, balloon-frame walls and mechanical penetrations allow outside air to penetrate buildings. Sometimes the problem is poorly installed insulation with too many voids, but more often, holes in the building are the real culprits.

Caulks and weatherstripping can help plug small holes (sidebar p. 19), but this is like using Band-Aids to treat major wounds. The total area of the holes I'm talking about is measured in

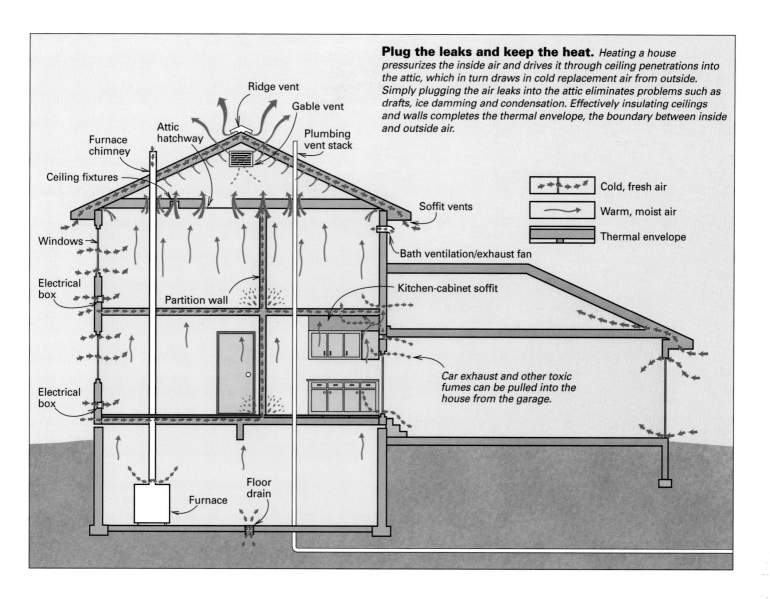

Plug the leaks and keep the heat. *Heating a house pressurizes the inside air and drives it through ceiling penetrations into the attic, which in turn draws in cold replacement air from outside. Simply plugging the air leaks into the attic eliminates problems such as drafts, ice damming and condensation. Effectively insulating ceilings and walls completes the thermal envelope, the boundary between inside and outside air.*

Ridge vent
Gable vent
Attic hatchway
Plumbing vent stack
Furnace chimney
Ceiling fixtures
Windows
Electrical box
Partition wall
Soffit vents
Bath ventilation/exhaust fan
Kitchen-cabinet soffit
Electrical box
Car exhaust and other toxic fumes can be pulled into the house from the garage.
Furnace
Floor drain

Cold, fresh air
Warm, moist air
Thermal envelope

square feet, not in square inches. Even so, these problems can now be fixed simply and economically, and buildings a century old can be routinely upgraded to higher performance levels than typical new homes. And the principles and methods are applicable to new construction.

Air movement in floors, walls and ceilings is bad—Air infiltration is the predominant heat-loss mechanism for most buildings (drawing above), so the primary goal of any weatherization effort should be to control air infiltration. Not all infiltration is bad; humans, pets, and furnaces and other combustion devices need a continuous supply of fresh outside air, and the air in most homes should be replaced (either naturally or mechanically) about six to eight times per day.

But relying on a home's air leaks is not a good way to provide fresh air. I've worked on buildings that have suffered as many as 30 air changes per day. At that rate, the conditioned air doesn't hang around long enough for the house's insulation to have much of an effect on keeping it in.

Air infiltration forces warm, moisture-laden air into cold, dry places. The buoyant nature of hot air drives it into every ceiling penetration, and if there are large holes, the house acts like a giant chimney, pulling cold fresh air in from below, heating it and pouring it into the attic. Loose attic hatches, large cutouts for plumbing vents, exposed beams and recessed lights are perfect "chimney flues" for these air currents (top photo, facing page).

When moist air contacts a cold surface in the wall or attic, water vapor condenses. If the building has a large reservoir of moisture—a wet cellar or an unvented bath, for example—terrible things can start to happen: Recessed lights drool; drywall stains and seam tape lifts; and exterior paint peels off soaked siding and trim. And once the sheathing hits 30% moisture content, mold and mildew can start growing, a condition carpenter ants and other bugs love. This chimney effect also causes a pressure drop in

the basement. Now the living areas are competing with the chimney flues for combustion gases. When the lift through the building overpowers the flues, backdrafting results.

Leaky return ducts in a forced hot-air system (bottom photo, facing page) can also vacuum up extra air from the basement and pressurize the living areas, driving conditioned air into the walls. As warm air is forced out, outside air rushes in to replace it. If the basement is tight, that air will come down the chimney, and potentially dangerous backdrafting will start again.

Combustion efficiency will drop, and dangerous pollutants from incomplete combustion, including carbon monoxide, can spill into the basement, be picked up by return ducts and be delivered efficiently to the rest of the house, a potentially life-threating situation.

Outside-air intrusion is another classic source of air movement in building cavities, blowing in through openings in walls and running the length of floors before exiting at the other end of the house. This cold outside air immediately comes in contact with warm interior surfaces, chilling them and causing moisture to condense. Ceiling corners are especially susceptible. Here mold can grow, and paint, ceiling texture and tape can peel off. Contrary to popular belief, most insulation doesn't block air intrusion (bottom photo).

Effective insulation and air-sealing take place at the thermal boundary—A thermally efficient building must have a well-defined boundary between indoors and outdoors. The holes and voids that allow outside-air intrusion are obvious breaks in the thermal boundary, but sometimes it takes a little head-scratching to figure out just where the boundary is. It's a waste of time and money to insulate an area that is actually outside the thermal boundary, so it's important to attack the right combination of floors, walls and ceilings to yield a complete thermal envelope.

For example, although attics and basements are usually thought of as being transitional areas between inside and outside, they really aren't. There is no "in between" in a properly weatherized house. I generally consider basements and crawlspaces to be inside the thermal envelope because it is difficult to isolate these areas from the living spaces above. Besides, combustion appliances always belong inside, where they operate more efficiently and can contribute Btus to the heated space.

On the other hand, vented attic spaces should always be outside. If the attic is used often, treat the access as an exterior door, and insulate the stairwell walls and under the stairs. When an attic is rarely used, a well-sealed foam hatch over the well is sufficient. I like to use surplus sections of stress-skin panels here. They are heavy enough to compress the gasket we place around the well, they are well-insulated, and the drywall is ready for paint.

Areas behind a kneewall can fall either inside or outside, depending on the use of the space. Because air infiltration can be a real problem here, special care should be taken to seal off the floor, kneewall and sloped ceiling from the outdoors. Air-sealing and insulating the rafters down to the bottom of the band joist brings this triangular storage space inside so that it can be used for easily accessible storage. If this space is inaccessible or unusable for storage, my favorite technique is to solidify the entire volume by packing it densely with cellulose, which air-seals it and insulates it at the same time. (For more on dense-packing cellulose, read on.)

Smaller holes and cracks can be filled with caulk or foam. After attic insulation is pulled away, holes can be found and filled. Here the author fills cracks in a plaster-and-lath ceiling.

Dirty fiberglass signals an air leak. As warm air pours through penetrations in the ceiling, dirt is filtered out by fiberglass batts, but the heated air goes into the attic. Holes for wiring should be filled with expanding-polyurethane foam to stop the loss of this air.

A blower door and careful investigation help to find the holes—I use pressure diagnostics to help direct my air-sealing efforts. Depressurizing the inside of a house with a blower door quickly reveals the most significant penetrations of the thermal boundary. But it doesn't take a blower door to find a lot of the major holes in the thermal envelope.

Under natural conditions, pressures are always higher at the ceiling than at the windows. Although wind and mechanically induced pressures are sometimes stronger, hot air applies constant pressure upward toward the ceiling and the attic. As a consequence, ceiling bypasses, or holes in the thermal boundary, generate more significant natural infiltration through the heating season than do window leaks.

This doesn't mean that door and window weatherstripping isn't cost-effective, but it does

Bottom photo this page: Fred Lugano

Tightening up doors and windows

Unless the glazing is actually broken out, I've found that doors and windows are the most expensive and least productive areas for thermal renovations. A cold house may have between 10 sq. ft. to over 100 sq. ft. of air intrusions and leaks. By contrast, a rattling and ill-fitting door or window won't have more than a few square inches of leakage.

If you do choose to weatherstrip windows, avoid cheap, quick-fix products. They aren't durable or effective enough to warrant the effort to install them. I like Randall's products (Randall Manufacturing Corp. Inc., 200 Sylvan Ave., Newark, N. J. 07104; 201-484-7600), which are available in many hardware stores and feature both resilient vinyl bulbs and Q-LON bulbs.

The small, hollow vinyl bulb is easily compressed and conforms well to irregular shapes, which makes it a good choice for attic hatches. I like the Q-LON bulb, often original equipment on new doors, for its ability to conform to large bows and warps by folding and compressing. Metal carriers work well in tight spaces or with metal jambs.

I've also used products from Resource Conservation Technology (2633 N. Calvert St., Baltimore, Md. 21218; 410-366-1146) with good success.

Dealing with old doors and windows—Old doors or windows have character and are usually worth fixing up. Simply adjusting the stops will tighten up a rattling double-hung window. Products such as pulley covers (Anderson Pulley Seal, 5158 Bloomington Ave. S., Minneapolis, Minn. 55417; 612-827-1117), retrofit vinyl jamb liners and side-mounted sash locks (H. B. Ives, 62 Barnes Park N., Wallingford, Conn. 06492; 203-265-1571) can also help to make a tight seal.

To make a seal, the bulb must compress slightly when the door is closed, and I've found that the Door-Tite ratcheting striker plate (Trion Co., P. O. Box 110358, Carrollton, Texas 75011; 800-532-9995) can help folks who have trouble generating enough force to make the door latch completely. This plate has a series of

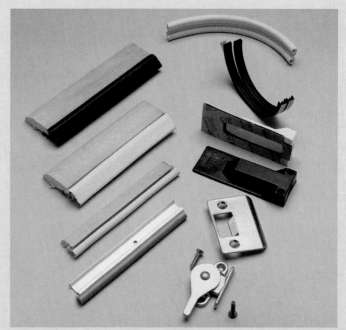

Tools of the trade. Counterclockwise from top left: Brown and white Q-LON weatherstripping; vinyl bulbs mounted in wood and metal carriers; side-mounted sash lock; ratcheting strike plate; brown and white pulley covers with adhesive backing; weatherstrip tape; P-profile EPDM tape.

stepped flats instead of one surface for the bolt to catch on.

And what about caulk?—I think life is too short to justify the time spent filling a square foot of $\frac{1}{16}$-in. cracks. If you use caulk, use it sparingly. I carry a dull putty knife and a damp towel to cut corners tightly and to wipe surfaces flush. Rutland makes several paintable, acrylic co-polymer caulks that adhere aggressively and tool cleanly (Rutland Products, P. O. Box 340, Rutland, Vt. 05702-0340; 800-544-1307). Rutland 500 RTV is the standard when you're air-sealing chimneys to sheet metal.—F. L.

---

mean that most doors and windows don't need replacement. There are reasons to replace windows, but unless there is glass missing or a large gap between sash and jamb, thermal performance is not a compelling one. There are better places to spend energy-conservation dollars.

Another place to concentrate on is the common wall between the house and its attached garage, if there is one. Air leaks here always have the potential to vacuum car exhaust, solvent and weed-killer fumes, and fuel gases into the living space, so this is a spot that requires a NASA-grade air-seal. Obvious holes are usually easy to find and fix in the open framing. Caulking framing and sheathing joints down to and along the foundation makes a big difference.

Basements and crawlspaces should also get a thorough inspection. Musty odors are a sure sign that moisture and cold air are mixing and that wood is under attack. Crawlspaces are usually

built to save money, and difficult access is often a reliable indicator of potentially significant building defects. We often have to saw our way into crawlspaces, where we can find bare soil, open concrete-block cores, no insulation, no sill seal, empty whiskey bottles, mold, decay, and lots of insect and animal debris.

Blocking moisture in the form of water vapor from the soil with 6-mil poly is an important first step to air-sealing here. Cover the ground completely, overlap seams if there are any, and lap the poly right up onto the foundation wall. Then the foundation, sills and band joists can be air-sealed and insulated with sheets of rigid foam and plenty of caulk. It's tough to do perfect work in a tight space. Sometimes it's possible to work from the outside by applying rigid-foam panels or stuccoing the stonework.

With the house depressurized by the blower door, I feel for drafts with the back of my hand

and spray expanding-urethane foam into trouble spots. Spiders can also offer clues; they always hang their webs in a draft. If the combustion devices have separate fresh-air supplies, foundation walls should be sealed as tightly as possible all the way to the ground, including foundation vents. Although often required by code, foundation vents allow crawlspaces to load up with moisture in warm months and allow cold air to circulate freely through the thermal envelope in the cold months. If moisture can be prevented from entering this space, then it doesn't need to be vented out.

It's important always to work at the boundary of the thermal envelope. Often during a blower-door test, an air leak to an electrical outlet, radiator pipe or wainscoting will show up in the middle of the house. But it won't do any good to stop the airflow there because the air will just find another exit point. Leave these interior

Ceiling fixtures and plumbing can be trouble spots. Bath fans are a good candidate for caulking, but the irregular hole around the vent stack is better sealed with foam. Insulated ductwork will keep the warm, moist exhaust air from cooling and condensing before exiting.

Larger air passages can be blocked with insulation-filled poly bags. Rafter bays are blocked with bagged fiberglass insulation prior to being insulated from above with blown cellulose. Later the plaster-and-lath walls will be insulated as well.

Blown cellulose makes a good insulation blanket. An alternative to fiberglass batts, cellulose easily flows over and around framing and into cavities. Eliminating gaps or voids helps to prevent cold air from washing through the thermal envelope.

holes alone and track down where the airflow actually enters the envelope. Air can travel long distances through floor bays and interior partitions in conventionally insulated homes. Air-sealing away from the envelope only redirects the airflow to another hole.

The most effective air-sealing is done in the attic—Insulation must trap still air; it won't work with air blowing through it. Current residential-insulation practice often ignores this fact. Many homes have vented attics that are actually inside the thermal envelope because the ceiling has so many penetrations. Instead of being trapped by the insulation, warm air pours right up through and into the attic. Everything looks fine when the holes are covered with a blanket of insulation, but when melting snow shows the rafter pattern on the shingles, it becomes obvious that the thermal boundary is really the roof

deck. The 12 in. of insulation in the ceiling is yielding an R-value of close to 0.

This isn't the time to add more useless insulation or ugly vents or an ice-dam membrane under the shingles. The best way to correct the problem is to dig through existing insulation and find and seal air leaks (top photo, p. 18).

Partition walls without top plates in older homes are often a major source of air leaks. Plumbing, wiring and chimney penetrations should be checked, and light beaming up into the attic from a ceiling fixture below is a sure sign of trouble.

I also look for blackened insulation (bottom photo, p. 18). As warm air finds a hole and jets through the insulation and into the attic, the dirt gets filtered out. Batt insulation wasn't designed to stop the loss of warm air from a building, but it does a good job of cleaning it.

Framing around chimneys should be sealed to the masonry with sheet metal and high-temperature silicone caulk. Mechanical penetrations are usually filled with nonexpanding-polyurethane foam from a gun (photo top left), while larger holes are best stuffed with fiberglass insulation wrapped in a poly bag (photo top right). We generally recycle our empty cellulose bags this way. For bigger holes, fasten down appropriately sized sheets of rigid foam, metal or ¼-in. oriented strand board and caulk the edges.

There are sometimes areas where it is difficult to rebuild a solid, continuous thermal envelope. For example, suspended ceilings are usually real trouble. When batts are laid over the grid, the assembly behaves like an open skylight, and air flows up through all the openings. In cases such as this, dig into the building until something solid and patchable can be found.

Dense-packed cellulose fills the gaps and stops the leaks—Once the flow of warm, moist, indoor air is cut off from the attic, the space can be prepared for adequate insulation. Because of the difficulty of installing fiberglass batts properly in an attic, I like to use blown cellulose. Potential ignition sources such as unrated recessed lights and chimneys should be dammed with sheet metal in order to keep them from contact with insulation. Hatches and soffit vents can be dammed with scrap lumber or plywood. This typically involves cutting and fitting a 10-in. or 12-in. deep box, or well, around framing so that it surrounds whatever needs to be dammed and keeps loose insulation out. Flagging electrical-junction boxes is a nice touch that you or your electrician will appreciate when it comes time to find them again. Finally, cellulose can be blown in at low pressure and low speed on top of the existing material to yield an honest R-40 (bottom photo). Treating open cavities is the easy part of weatherization.

Insulating and air-sealing closed cavities is more difficult. In old homes with plaster walls and board sheathing, you can't simply caulk or foam every seam. Fortunately, dense-packing cellulose into closed cavities provides a cheap method of insulating and air-sealing in one step; it is so effective that it has become a cornerstone of weatherization practice. Instead of being fluffed in, the insulation can be crammed in at twice the conventional density. At 3.5 lb. per cu. ft., it becomes an air-sealing medium too dense for wind to penetrate, while at the same time maintaining its R-value.

To gain access to the cavities, we drill a 2½-in. access hole from either the inside or the outside (photo right). When we can, we work outside because blowing cellulose is a dusty job. Some contractors drill right through the siding, but usually enough siding can be removed to gain direct access to the sheathing. Later, after the holes are drilled and the cellulose is blown in, the siding can be reinstalled.

A vinyl tube is snaked through until it bumps the end of the cavity. The tube acts as a vertical probe, and if it doesn't extend to either end of the bay, another hole will have to be drilled above or below the blockage. We blow in a lean mixture of air and cellulose at high speed. As the bay pressurizes, the fine cellulose particles flow into every crack.

When the pack becomes airtight, it stalls the flow in the hose, so we pull the tube back until it finds more loose fill. The completed bay is now solid, insulated to R-3.8 per in.; fire-, insect- and rodent-resistant; and air-sealed. When the cellulose is dense-packed so tightly that it stops the flow of air from the blower, it will also stop any wind pressure that nature can exert.

We use this method in walls, in cantilevered floors, under attic stairs and in odd triangular spaces behind kneewalls. Dense-pack is also an effective method for insulating cathedral ceilings, the inaccessible thin edges of attics under shed roofs and inaccessible joist and rafter bays. I think that properly installed cellulose is the finest insulation technique for new construction, too, but its versatility and ability to fill and to air-seal voids in the wall cavities of old houses makes it indispensable in effective weatherization. Together with attic air-sealing, a dense-packed envelope will generally cut natural air infiltration in half before doors, windows and basement are even touched.

Although dense-packed cellulose won't bring R-19 levels to 2x4 walls, it will still reduce infiltration in ancient buildings to minimal rates. Because most heat loss is caused by air changes anyway, these beautiful and invisibly updated period houses can now perform at higher comfort and efficiency levels than conventionally insulated new ones.

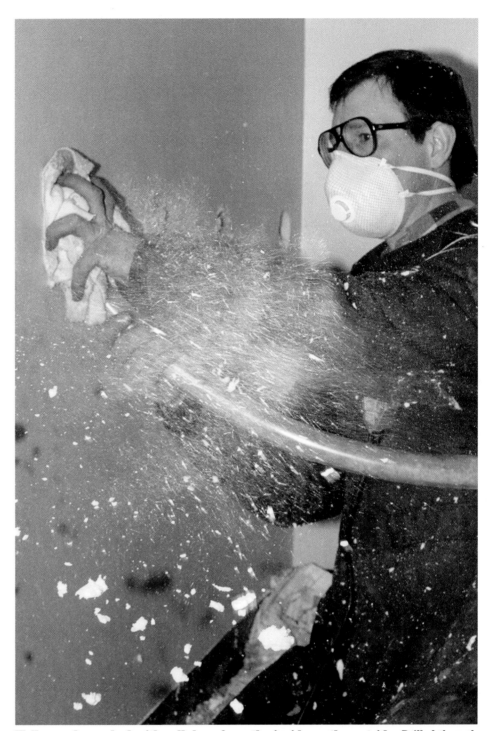

Walls can be packed with cellulose from the inside or the outside. Drilled through either the exterior sheathing or the interior finished wall, 2½-in. holes provide access to the stud bays. A rag placed over the opening while it is being tubed helps to control the dust.

Weatherization is tough business—Insulating a building is physical, dirty work that can take you into tight, uncomfortable spaces. I don't know of anyone who likes working in a confined space at 140°F wearing a respirator.

And what is the payoff for these weatherization efforts? Even in times of relatively stable, low fuel prices, a 20% to 30% return on investment in fuel savings is the norm. Even better are the long-term maintenance issues, such as peeling paint and ice damming, that effective weatherization helps to solve. But best of all is the increased level of comfort for the home's residents: No longer does an old house—or even a new house—have to be cold, drafty and difficult to heat. □

Fred Lugano owns Lake Construction and is a weatherization contractor in Charlotte, Vermont. Photos by Andrew Wormer, except where noted.

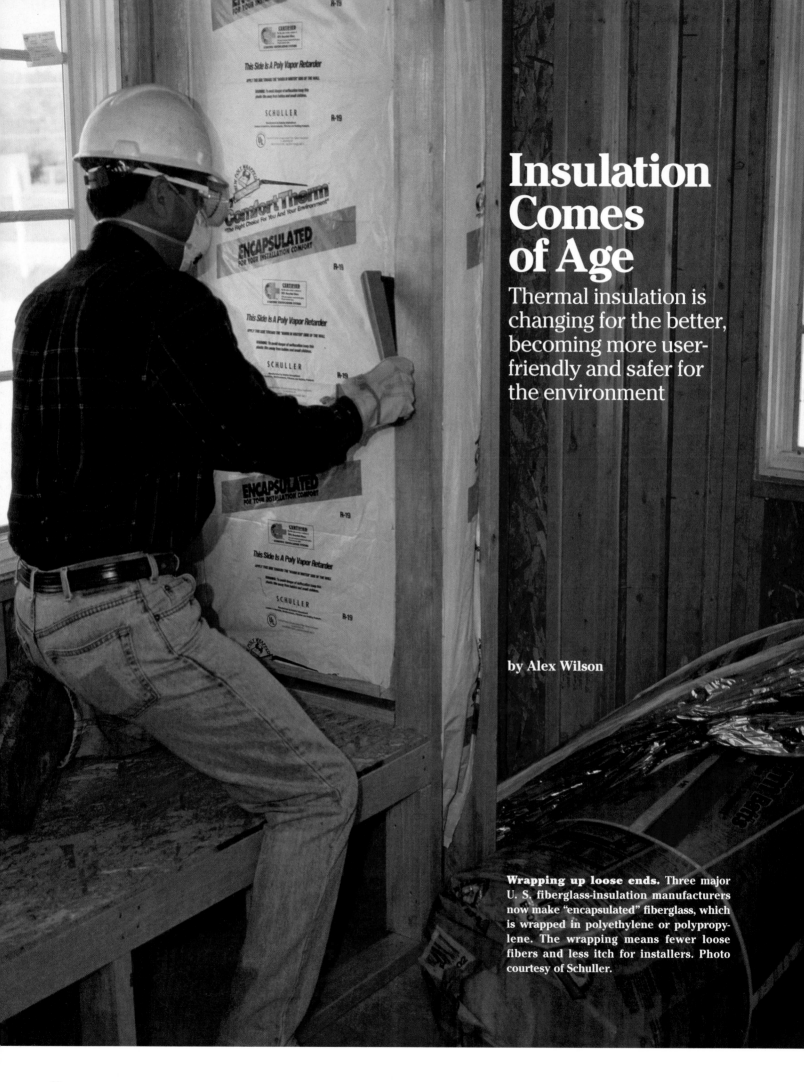

Insulation Comes of Age

Thermal insulation is changing for the better, becoming more user-friendly and safer for the environment

by Alex Wilson

Wrapping up loose ends. Three major U. S. fiberglass-insulation manufacturers now make "encapsulated" fiberglass, which is wrapped in polyethylene or polypropylene. The wrapping means fewer loose fibers and less itch for installers. Photo courtesy of Schuller.

I wasn't a job for the squeamish. Just to squeeze beneath the massive log joists of my old cape, I first had to dig my way through two centuries of bat droppings and crawlspace dirt. Then I had to drag in boards of 2-in. polystyrene, nail them between the joists and seal the polystyrene to the joists with the kind of foam sealant that works only when the can is upside down.

The thought occurred to me—there in the dark dampness, my back on a bed of filth, my face up against the spiderwebs—that it would be a lot easier just to keep burning cordwood and enduring the cold drafts. I let the thought pass and kept working.

To this day, that insulation project lives vividly in my mind as the worst job I've ever undertaken. Yet it was also one of the most satisfying. One day of filthy, cramped labor cut our wood consumption by a third and reduced the home's 30°F floor-to-ceiling temperature difference to an acceptable 10°F. It also demonstrated to me what a difference insulation can make. In one day I changed the thermal dynamics of the house more dramatically than at any other time in its 200 years. I used the right type of insulation and an understanding of issues such as moisture flow to ensure a continuing long life for the joists and floor of the historic house.

That was a dozen years ago. In the relatively short time since, thermal insulation has changed a lot. Today, those changes are coming fast and being driven by a variety of causes. Health and safety concerns, better understanding of building science, changes in regulations, environmental awareness and economics all contribute to the rapid evolution of the ways we protect our homes from temperature extremes. In the six years since *Fine Homebuilding* last tackled this broad issue (*FHB* #56, pp. 36-42), new materials have become available, and existing products have gone through redesigns.

I'll walk you through the changes and look at the insulation industry material by material, highlighting what's new both with markets and with products, and along the way I'll revisit some old standards (chart p. 27).

Fiberglass continues to lead the batt pack—The largest share of the cavity-fill insulation market belongs to fiberglass. Although this insulation material has been around for decades, even fiberglass has had changes recently.

The biggest issue driving change in the fiberglass-insulation industry is concern over health and safety. Although hotly disputed by the fiber-

Wet-spray cellulose is trimmed after installation. After a cavity is filled with a cellulose spray, the cellulose is rolled to remove any excess, which is then thrown back into the hopper for reuse. (A mask should be worn during installation of cellulose.) Photo courtesy of Green-Stone Industries.

Blown-in cellulose for new construction. Before the drywall goes up, cellulose insulation can be blown into cavities through holes in a taut layer of reinforced plastic installed over the studs. Photo courtesy of ParPack.

glass industry, some researchers claim that breathable glass fibers can cause cancer, which led to a requirement by the federal Occupational Safety and Health Administration (OSHA) that fiberglass and mineral-wool insulation packaging carry warning labels.

Manufacturers found that the best way to prevent fibers from escaping into the air was to contain them. Consequently, all three major manufacturers of fiberglass insulation—Owens-Corning, Schuller and CertainTeed—addressed the perceived risk by introducing "encapsulated" fiberglass-batt products for the do-it-yourself markets (photo facing page). Batts are wrapped in a perforated polyethylene or, in the case of CertainTeed's (800-433-0922) new product, a nonwoven polypropylene fabric.

BIBS encapsulates a different way—Abiff Manufacturing (800-852-1369) developed the Blow-In-Blanket System (BIBS) for fiberglass and other fiber insulation materials to achieve a bet-

ter seal around wires and pipes. The BIBS system is marketed nationally by Ark-Seal International (800-525-8992). With BIBS, a mesh netting is stapled over the faces of frame walls and ceiling cavities, and a mixture of the fiber insulation and binders is blown in the holes. The binder ensures a good seal and prevents settling.

High density gives more R per inch—The three major fiberglass manufacturers—and the smaller Guardian Industries (800-562-0863)—also introduced higher-density, higher-R-value batts. Fibers are tightly packed together; airspaces are smaller and more effective at preventing air circulation. So you get more insulation for a given thickness. Now there are 3½-in. R-15 batts for 2x4 walls; 5½-in. R-21 batts for 2x6 walls; 8¼-in. or 8½-in. R-30 batts for 2x10 rafters where an airspace is required; and 10¼-in. R-38 batts for 2x12 rafters where an airspace is required.

The new high-density fiberglass costs about 20% more, however, depending on the product

An idea America might cotton to. Cotton insulation comes loose fill or in batts and is billed as providing R-values comparable to fiberglass. Photo courtesy of Greenwood Cotton.

to be used. You can accomplish pretty much the same effect by cramming a thicker standard batt into an insulation cavity. A standard R-19 6-in. batt stuffed into a 2x4 wall cavity, for example, will provide R-14—almost as much as the high-density 3½-in. batt. But this approach also costs more and doesn't work as well with insulated cathedral ceilings where you need an airspace. Here, the new high-density batts can be used without a vent spacer, while thicker batts require a vent spacer to maintain the airspace.

They've finally done something about those irritating fibers—The biggest news with fiberglass insulation in years is Miraflex (photo above right), a new type of fiber from Owens-Corning (800-752-1986). Miraflex rolled out nationally in the company's R-25 Pink Plus product in September 1995. While more expensive than standard fiberglass, it is the same price as Owens-Corning's other encapsulated fiberglass.

Miraflex is the fusion of two different types of glass. Because the two types of glass expand at different rates as they cool, the fibers curl and twist together, which means they can be held together in batts without the need for a phenol formaldehyde binder. The result is a fiberglass material that's soft to the touch—not scratchy like ordinary fiberglass—and easy on the skin.

Perhaps even more significant, Miraflex fibers are more flexible and have greater tensile strength than conventional glass fibers, so they don't break off and become airborne. As a result, Owens-Corning omits the cancer-warning labels on Miraflex.

Mineral wool costs more than fiberglass, but it can take the heat—Mineral wool once held the largest share of the insulation market but lost most of it to fiberglass during the 1960s and 1970s. Mineral wool is similar to fiberglass, except that natural stone or iron-ore blast-furnace slag is the raw material for the insulation. If stone is used (typically basalt or diabase), the product usually is called rock wool. If iron-ore blast-furnace slag is used, the product is usually called slag wool.

Mineral wool often is used in commercial applications because it can withstand higher temperatures than fiberglass and has better acoustic insulation properties. However, it's a lot heavier, tends to be more expensive and generates the same health concerns as fiberglass.

The European company Roxul (800-265-6878), which has a manufacturing plant in Canada, uses a 50-50 mixture of stone and slag in its mineral wool. Roxul is the only mineral-wool company known to be producing batt insulation.

Called Flexibatt, the Roxul product has a flexible edge that allows it to be compressed and squeezed into a joist bay, where it is supposed to stay without the need for fasteners or holders. Some industry observers believe Roxul may significantly expand its market in North America.

All the news that's fit to be fiberized—Cellulose insulation is ground-up newspaper that's treated with fire retardants and used as loose-fill attic and wall insulation. There's not a lot new with cellulose, except for the cost of the raw material. The price of newsprint recently increased a great deal, so manufacturers are switching from the old "hammer-mill" process of chopping up newspaper to a newer "fiberizing" process for the material.

Installed densities as low as 1.3 lb. per cubic foot (pcf) are possible with fiberized cellulose, according to some manufacturers, although values of 1.5 pcf are more typical. (For comparison, the density of fiberglass is usually 0.5 pcf to 0.75 pcf.) With fiberized cellulose, material savings of 25% are typical, and the R-value per inch increases slightly, according to some manufacturers. To reduce settling of cellulose attic insulation—especially with lower-density material—some installers are going to a "stabilized" installation process. A small amount of moisture-

A house in sheep's clothing. Shear factor takes on a new meaning when wool insulation is installed in a house, such as this New Zealand house that's being fitted with batts of Woolhouse insulation. Photo courtesy of Woolhouse.

activated acrylic binder and water is added to the cellulose during installation to bind the fibers together and to reduce settling greatly.

Cellulose can be blown into walls at a high enough density to prevent settling (typically 3 pcf to 3.5 pcf); mixed with water (and sometimes a small amount of water-activated acrylic binder) as it's sprayed into walls in the "wet spray" process (top photo, p. 23); or blown into cavities through holes in a taut layer of reinforced plastic installed over the studs (bottom photo, p. 23).

The fire safety of cellulose has been challenged for years, largely by the fiberglass industry. Because wood fiber—the basic raw material in cellulose insulation—burns and glass does not, fire-safety concerns are justified. However, most evidence suggests that properly treated cellulose poses no greater fire risk than fiberglass. In fact, because cellulose blocks airflow better than fiberglass, cellulose may be more effective at stopping fire spread.

Cotton insulation fits like a pair of old jeans—Cotton insulation was first introduced about five years ago under the brand name Insulcot, which has since sold the manufacturing rights to Greenwood Industries in Roswell, Georgia (404-998-6888). That company introduced

Greenwood Cotton Insulation in 1995. Greenwood is 95% post-industrial recycled textile-mill waste. The insulation is 75% cotton and 25% polyester (photo left, facing page).

Available in both kraft-faced batts and loose-fill forms, the insulation contains flame retardants similar to those used in textiles. There are no formaldehyde binders.

Greenwood Cotton is now marketed primarily in the southeast and mid-Atlantic states (within about a 400-mile radius of the manufacturing plant in Georgia), although Greenwood's Kirk Villar said a lot of special orders have come from California. The product is popular among healthy-home builders who are concerned about possible health and safety problems with fiberglass and other types of insulation.

My experience with preproduction samples of the product was mixed. It was wonderful not ending the work with itchy arms and eyes, as I've experienced with fiberglass. But the product seemed quite different from fiberglass. The batts didn't immediately spring back, or loft, to their full thickness. After I slit the bags open, the batts just laid there. The company has since modified the density and fiber composition, which apparently improved the lofting. The batts are also difficult to cut, another concern being addressed by Greenwood.

A warm blanket for the house—If insulation can be made out of cotton, how about wool? That's exactly what the New Zealand company Woolhouse International Ltd. (011-64-9-486-7020) did in 1992 with Thermofleece Natural Wool Insulation (photo above). The company is looking into the possibility of building a North American manufacturing facility.

In the Thermofleece product, specially processed wool is mixed with other polymer materials to give it spring back, or loft. A boric-acid fire retardant is added to the New Zealand product, but U. S. testing has not been done.

There is at least one other wool-insulation manufacturer in New Zealand, and there are two in Germany. Paul Novak of Environmental Construction Outfitters in New York City says he has looked into importing wool insulation, but that the cost was just too high to compete, even for specialty markets. A domestic manufacturing facility could change the picture.

Board-stock insulation still does the same job, but it's less harmful to the environment—The biggest change with extruded polystyrene (XPS) in recent years was the substitution of HCFC-142b for CFC-12. Dow Chemical and Amoco Foam Products completed this transition in 1990, the other manufacturers by the end of 1993. Although HCFCs are ozone-depleting substances, they are far less destructive than the CFCs.

Amofoam (800-241-4403) offers two product lines that contain recycled materials: Amofoam-RCY and Amofoam-RCX. Both products have 50% total recycled content: 25% post-industrial recycled polystyrene and 25% post-consumer recycled polystyrene. The RCY product has been on the market for several years. The RCX product, which is aimed at residential markets, is thinner (½ in. or ¾ in.) and is faced with polyethylene on both sides. It was introduced in 1995. Amofoam products are available only east of the Rockies, except by special order.

Expanded polystyrene (EPS, or beadboard) is produced by expanding beads of polystyrene in a mold, then slicing the block of molded foam into boards. Pentane, which does not deplete ozone, is the blowing agent.

Because EPS is the only rigid-foam insulation not produced with ozone-depleting substances, it has gained the favor of environmentally concerned builders and designers. Some manufacturing plants, particularly those in California, have pentane-collection systems to reduce plant emissions. In addition, several suppliers of polystyrene beads have introduced low-pentane beads that use about half of the pentane found in typical beads.

While most EPS board stock is 0.9 pcf to 1 pcf, densities up to 2 pcf are available from most manufacturers. At higher densities, compressive strength is greater, R-value is higher, and moisture resistance is better—as is needed below grade. EPS in density of 1.5 pcf to 2.0 pcf is commonly used in insulated foundation forms for below-grade applications. Cost is greater for the higher-density EPS.

AFM Corporation (800-255-0176) began incorporating a borate insect repellent into some of its EPS foam in 1990. The repellent is now standard in AFM's R-Control panels and available as an option in other EPS products, such as its Diamond Snap-Forms (bottom photo, p. 28). The repellent is added to address the concern that termites and carpenter ants often tunnel through foam-core panels or foam insulation on the outside of foundations and may use the foam as a way into a house. So far, the borate treatment seems to work successfully.

A final development with EPS is the appearance of polyethylene facings on some products to improve durability, both for roofing products and for foam sheathing (top photo, p. 28). With a more durable product, we are likely to begin seeing more EPS sold in building-supply yards, where it's rarely been stocked in the past.

Polyisocyanurate-foam insulation is widely used as an exterior insulative wall sheathing and as roof insulation. Most products are foil-faced,

Spray it on, saw it flat. Icynene, an open-cell polyisocyanurate insulation, is sprayed on like paint and within minutes expands 100% to fill all voids in a building cavity. Excess is trimmed away with a handsaw.

although other, more-rugged facings have appeared in recent years, such as polymer-coated glass fiber.

Polyiso foam used to be blown with CFC-11. The blowing agent that is being used now, HCFC-141b, is much less damaging to ozone. However, this one is among the worst of the HCFCs, relative to ozone depletion, and is slated to be phased out by 2003. Suppliers of chemical blowing agents are hard at work on so-called third-generation compounds that have no impact on ozone.

Water is replacing hydrocarbons in polyurethane insulation—Chemically similar to polyisocyanurate-foam board stock, spray polyurethane is used in cavities or applied over surfaces to be insulated. The use of spray-polyurethane insulation dropped off significantly as the cost of the CFC-11 blowing agent increased. But now that the shift to HCFC-141b is complete (as of January 1, 1994) and as the cost of wet-spray cellulose insulation increases, use of polyurethane is picking up again.

The most significant developments in this area are products foamed with carbon dioxide or water in place of CFCs or HCFCs. The leading product in this area is Icynene (800-758-7325), a carbon-dioxide-blown, open-cell modified polyurethane foam developed in Canada about 10 years ago (photo above). The company recently expanded into the United States and now has 60 licensed installers in 25 states.

Icynene is sprayed into open wall cavities in a very thin layer, almost like spray paint, and it expands immediately. In a few seconds Icynene expands 100-fold, filling the cavity. Because the expansion is so rapid and so great, installers often overfill cavities and have to go back and cut off the excess using a handsaw.

As an open-cell foam, Icynene is spongy to the touch. It adheres extremely well to most surfaces and is effective at providing an air seal. Marketed in Canada as InSealation, Icynene expands to a density of approximately 0.5 pcf, or roughly one-fourth the density of standard spray polyurethane or polyiso board stock. Its insulating value is approximately R-3.6 per inch.

Icynene recently introduced a second product designed for installation into closed cavities. A carefully measured volume of the unexpanded foam is poured into the cavity, where it expands from bottom to top to fill the cavity. This product insulates to R-4 per inch. Also, Icynene

Insulation specs

Type of insulation	Average R-value per inch	Available thickness	Where used	How installed	Resistance to:** Water absorption	Moisture damage	Direct sun	Fire	Temperature*
Batts, rolls									
Fiberglass	3.2	1 in.-13 in.	Wall, floor, ceiling	Fitted between studs, joists and rafters	2	1	1	2	180°F
Mineral wool	3.2	3 in.-8 in.	Wall, floor, ceiling	Fitted between studs, joists and rafters	2	1	1	1	500°F+
Cotton	3.2	3½ in.-12 in.	Wall, floor ceiling	Fitted between studs, joists and rafters	4	3	2	2	N/A
Loose, blown, poured									
Fiberglass	2.2	Variable	Ceiling, wall	Poured and fluffed, or blown	2	1	1	2	180°F
Rock wool	3.1	Variable	Ceiling, wall	Poured and fluffed, or blown	2	1	1	1	500°F+
Cellulose	3.2	Variable	Ceiling, wall	Blown by machine	4	4	2	3	180°
Cotton	3.2	Variable	Ceiling, wall	Poured and fluffed, or blown	4	3	2	2	N/A
Rigid board									
Expanded polystyrene (EPS)	4.0	¼ in.-10 in.	Wall, ceiling, roof	Glued, nailed	3	2	4	4	165°F
Extruded polystyrene (XPS)	5.0	¾ in.-2 in.	Foundations, subslab, wall, ceiling, roof	Glued, nailed	1	1	4	4	165°F
Polyiso-cyanurate	6.5	½ in.-4 in.	Wall, ceiling, roof	Glued, nailed	3	2	4	4	200°F
Rigid fiberglass	4.4	1 in.-3 in.	Wall, roof, ceiling and foundation	Glued, nailed	2	1	1	2	180°F
Contractor-applied									
Wet-spray cellulose	3.5	Variable	Wall	Sprayed into open cavities	4	3	2	3	165°F
Blown fiber with binder (BIBS)***	3.5-4.0	Variable	Wall, ceiling	Blown dry into cavities faced with mesh screening	**	**	**	**	165°F
Polyurethane	6.2	Variable	Wall, ceiling, roof	Foamed into open cavities	1	1	4	4	165°F
Open-cell polyurethane	3.6****	Variable	Wall, ceiling	Foamed into open or closed cavities	3	2	N/A	3	165°F
Phenolic foam	4.8	Variable	Wall	Foamed into closed cavities	2	2	N/A	3	212°F
Magnesium silicate	3.9	Variable	Wall	Foamed into open cavities	3	2	N/A	1	500°F+

*Maximum temperature insulation is rated to withstand.

***Properties depend on fiber insulation used.

**1=Excellent; 2=Good; 3=Fair; 4=Poor.

****4.0 for closed-cavity installation.

A close-up look at innovations in batt insulation

"Itch-free" Miraflex fibers won't get into your lungs.

Encapsulated fiberglass is wrapped up to stay put.

New cotton insulation is made from cotton-mill waste.

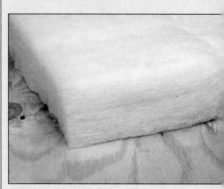

From New Zealand come wool insulation blankets.

Structural panels add strength to R-value. Polystyrene foam insulation is fixed to oriented strand board and used as sheathing. These R-40 roofing panels made by Insulspan are covered with 7/16-in. OSB on both sides. Photo courtesy of Insulspan Inc.

An insulating concrete form that resists termites. AFM Corporation adds borate to its expanded polystyrene Diamond Snap-Forms to help prevent invasion of concrete foundations by termites and carpenter ants. Photo courtesy of AFM Corporation.

isn't the only non-ozone-depleting spray foam. Resin Technology Company (909-947-7224) of Ontario, California, produces both open-cell and closed-cell water-blown polyurethanes and may begin marketing the foams for building insulation. FoamTech of North Thetford, Vermont (802-333-4333), produces SuperGreen, a non-ozone-depleting HFC-blown polyurethane foam. Introduced in 1993, this product costs about 10% more than conventional polyurethane, but it insulates just as well. It's used both as a building insulation and as an air seal for cracks and gaps.

Other spray-ins are abandoning CFCs, too—Spray-foam sealants, which are sold in aerosol cans or larger canisters, also have undergone significant change. These products are either one-part or two-part polyurethane foams. Like spray polyurethane, they have abandoned use of CFC-11, but several different propellants and blowing agents are now being used. Most domestic products are using HCFC-22 as the blowing agent, the same chemical used in home air conditioners.

The German product PurFill, distributed in this country by Todol Products (508-651-3818) of Nat-

A glimpse into the future

Insulating with gas-filled pillows. Batts filled with low-conductivity gases may be used in attics to achieve high R-values at lower cost. Photo courtesy of Lawrence Berkeley Laboratory.

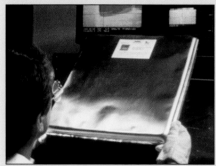

Would you believe R-75 per inch? High-insulation-value vacuum panels, such as this one developed by Owens-Corning, may eventually be used in ceilings. Photo courtesy of Owens-Corning.

ick, Massachusetts, switched from HCFC-142b to non-ozone-depleting HFC-134a in 1995. (HCFCs are being phased out in Europe much more quickly than in the United States.) Manufacturers here are also likely to shift to the HFC blowing agent. Insta-Foam already sells one product that uses only HFC-134a. Another approach is to use hydrocarbons, such as isobutane and propane, as the blowing agents. Convenience Products (800-325-6180) makes several such foam sealants under the Touch 'n Foam product name. While much more flammable than HFC-foamed products, the hydrocarbons are ozone-safe and, unlike HFCs, are not greenhouse gases that contribute to global warming.

Tripolymer Foam, made by C. P. Chemical (914-428-2517) of White Plains, New York, is foamed with compressed air, so it doesn't result in ozone depletion. Tripolymer is typically used for filling closed-wall cavities and masonry-block cores, but it can also be surface-sprayed into open cavities. The product insulates to about R-4.8 per inch but experiences shrinkage of 1½% to 3%, which reduces overall R-values. It can be installed only by licensed insulation contractors using the proper equipment.

Magnesium silicate oozes in like shaving cream—Air Krete (301-898-7848), an inorganic, foamed magnesium-oxide cement, is unique in the insulation industry. It's one of the few insulation materials people with acute chemical sensitivity can tolerate, which makes it popular among healthy-home builders. First commercialized in the late 1970s, the insulation is highly stable with virtually no shrinkage after setting, it is noncombustible, and it is free of virtually all volatile organic compounds (VOCs).

During installation, Air Krete is the consistency of shaving cream. It can be installed from the top of cavity walls, through holes or gaps in the drywall, or through screening attached to the interior face of framing. The cured foam has a typical density of 2 pcf and insulates to R-3.9 per inch. It is friable, so it must be protected within a wall or ceiling cavity.

There are currently about 50 installers of Air Krete in the United States. Like other foam-in-place insulation materials, Air Krete requires specialized equipment and training to install. The price of the installation depends on the size of the project and the distance the installer has to travel, but Air Krete is generally considered a premium product.

In the right place, radiant insulation makes sense—In the past 15 years, a number of products that work by blocking radiant heat flow have come onto the market. A few companies marketed their products with grossly exaggerated performance claims, which gave the industry a bit of a black eye.

There are really three distinct types of products that work by reducing radiant heat flow: radiant barriers, reflective insulation and radiant-control coatings. A radiant barrier is a foil or foil-coated plastic film that is installed next to an airspace. Reflective insulation is any board-stock, batt or bubble-pack insulation with a reflective component. Bubble-pack products have an airspace next to the radiant surface by function of their design. Board-stock and batt insulation with reflective facings only block radiant-heat flow if the reflective surface is installed next to an airspace. Finally, radiant coatings are applied to roofs or exterior-wall surfaces to reflect heat from sunlight.

In the right applications, radiant-insulation systems make a lot of sense, but they are rarely good substitutes for conventional insulation. The most common uses for radiant barriers are in unheated attics that are beneath the roof sheathing. In these applications, unwanted summertime-heat gain can be significantly reduced. This benefit of radiant barriers will only be achieved if there is little or no attic insulation. If attics are properly insulated, adding a radiant barrier is rarely cost-effective.

Beyond the blue horizon—While not yet available for building applications in the United States, a number of products are in the works that you may see in the not-too-distant future.

Among the most exciting insulation products to come along in many years is an R-75-per-inch vacuum panel now being produced by Owens-Corning (photo bottom left). Introduced in 1995, the Aura panel is a 1-in. thick layer of rigid fiberglass surrounded by an airtight skin of stainless steel. After the edges are sealed, a vacuum is drawn in the panel (air is removed), and a high R-value is achieved.

Aura insulation is currently being used to insulate freezer compartments of certain models of Whirlpool refrigerators—the 1-in. panels replace 2 in. to 4 in. of ozone-depleting polyurethane foam and provide far-superior insulation. In typical dimensions for refrigerators, 1-in. Aura panels have average insulating values of R-30 to R-40 (including edge losses). Whether this technology will ever make it into house-building applications is doubtful given the high cost and risk of puncture.

However, a similar vacuum technology that was developed at Oak Ridge National Laboratories is being tested in mobile-home ceilings. Researcher Ken Wilkes said the vacuum panels being used in the test contain a silica powder rather than rigid fiberglass. The panels yield R-values up to R-30 per inch, according to Wilkes, which makes them ideal for the shallow ceilings of mobile homes.

Researchers at Lawrence Berkeley Laboratory have been working for a number of years on an insulation idea borrowed from windows. Researcher Dariush Arasteh is pursuing the idea of putting low-conductivity gases, such as argon and krypton, into multilayer plastic pillow materials (photo top left). Because these gases have lower conductivity than air, they provide higher R-values.

Abiff Manufacturing of Denver, Colorado, is developing a new type of insulation made of fly ash, a waste product from coal-fired power plants. The company has developed a process to create little low-density clusters of fly ash that could be used in blow-in or loose-fill insulation applications. Inventor Henry Sperber hopes to have the product on the market sometime in 1996. Researchers at Oak Ridge National Laboratory and Denver University also are assisting in the effort. □

Alex Wilson is editor and publisher of Environmental Building News *in Brattleboro, Vermont (802-257-7300), a bimonthly newsletter on environmentally responsible design and construction. He is author of* Consumer Guide to Home Energy Savings *(American Council for an Energy-Efficient Economy, 1995).*

Preventing Ice Dams

Proper insulation and roof ventilation
can stop ice dams from forming,
prevent damage and lower
energy bills

by Paul Fisette

The call came on a sunny February afternoon: "The water line to my dishwasher burst, and I can't shut it off. Can you help me out?" I rushed across town to find a former client stuffing towels into the kick space below his kitchen cabinets. I quickly shut off the water to the dishwasher, but nothing happened. Although there wasn't a pipe anywhere near the leak, water still flowed. Puzzled, I made a brief investigation and found the source. The water running down the wall cavity was from an ice dam on the sunlit roof above.

This account is typical of dozens of ice-dam problems I've investigated as a builder, researcher and consultant. Although individual cases may look different and can result in different types of damage, all ice-dam situations have two things in common: They happen because melting snow pools behind dams of ice at the roof's edge and leaks into the house; and they are avoidable. The symptoms can be treated and the damage repaired, but the key to dealing with ice dams is preventing them in the first place.

Ice dams form when melted snow refreezes at roof edges—Everyone living in cold climates has seen the sparkling rows of ice that hang like stalactites along eaves. Most people, however, don't stop to understand what causes these ice dams until damage is done. Ice dams need three things to form: snow, heat to melt the snow and cold to refreeze the melted snow into solid ice (photo left). As little as 1 in. or 2 in. of snow accumulation on the roof can cause ice dams to form. Snow on the upper roof melts, runs under the blanket of snow to the roof's edge and refreezes into a dam of ice, which holds pools of more melted snow. This water eventually backs up under shingles and leaks into the building (top drawing, facing page).

The cause is no mystery. Heat leaking from living spaces below melts the snow, which trickles down to the colder edge of the roof and refreezes into a dam. Every inch of snow that accumulates on the roof insulates the roof deck a little more, trapping more indoor heat and melting the bottom layer of snow. Frigid outdoor temperatures guarantee a fast and deep freeze at the eaves.

There are a couple of reasons for the loss of heat from living spaces. First, on most homes rafters sit directly on top of exterior walls and leave little room for insulation between the top of the wall and the underside of the roof sheathing. Second, some builders aren't particularly fussy when it comes stopping the movement of warm indoor air into this critical area.

Avoid expensive damage by recognizing the signs of ice damming—Ice dams cause millions of dollars in damage every year. Much of the damage is apparent. We easily recognize water-stained ceilings; dislodged roof shingles; sagging, ice-filled gutters; peeling paint; and damaged plaster. So check your home carefully when you notice any of these signs.

Not all damage is as obvious as water stains on the ceiling, however, and some hidden damage can go unchecked. Insulation is one of the

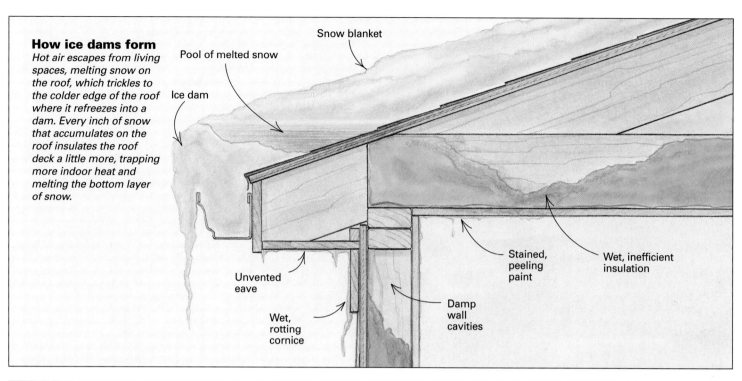

How ice dams form
Hot air escapes from living spaces, melting snow on the roof, which trickles to the colder edge of the roof where it refreezes into a dam. Every inch of snow that accumulates on the roof insulates the roof deck a little more, trapping more indoor heat and melting the bottom layer of snow.

Snow blanket

Pool of melted snow

Ice dam

Unvented eave

Wet, rotting cornice

Damp wall cavities

Stained, peeling paint

Wet, inefficient insulation

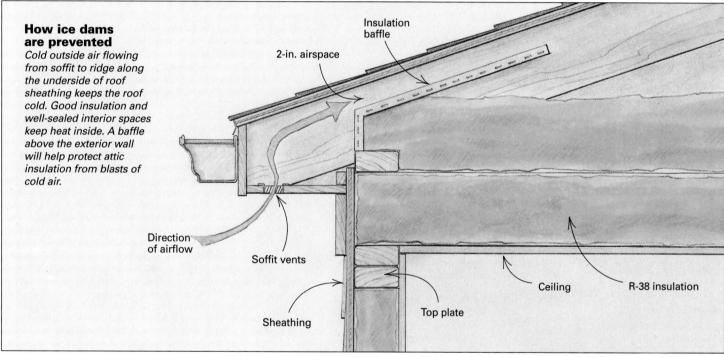

How ice dams are prevented
Cold outside air flowing from soffit to ridge along the underside of roof sheathing keeps the roof cold. Good insulation and well-sealed interior spaces keep heat inside. A baffle above the exterior wall will help protect attic insulation from blasts of cold air.

Insulation baffle

2-in. airspace

Direction of airflow

Soffit vents

Sheathing

Top plate

Ceiling

R-38 insulation

biggest hidden victims of leaks. Roof leaks dampen attic insulation, which in the short term loses some of its insulating ability. Over the long term, water-soaked insulation compresses so that, even after it dries, the insulation isn't as thick. Thinner insulation means lower R-values. As more heat leaks from living areas into the attic, it's more likely that ice dams will form and cause leaks. The more water that leaks through the roof, the wetter and more compressed the insulation becomes. Cellulose insulation is particularly vulnerable here. It's a dangerous cycle, and as a result, you pay more to heat and cool your house.

There's more: In the wall, water leaks soak the top layer of insulation and cause it to sag, leaving uninsulated voids at the top of the wall (top

drawing above). More heated air escapes. More important, moisture in the wall gets trapped between the exterior plywood sheathing and interior vapor barrier. The result is smelly, rotting wall cavities. Structural framing members can decay; metal fasteners can corrode; mold and mildew can form on wall surfaces as a result of elevated humidity levels; and exterior and interior paint can blister and peel.

Peeling wall paint deserves special attention because its cause may be difficult to recognize. It's unlikely that interior or exterior wall paint will blister or peel while ice dams are present. Paint peels long after ice and all signs of a roof leak have evaporated. The message is simple. Investigate even when there doesn't appear to be a

leak. Look at the underside of the roof sheathing and roof trim to make sure they're not wet. Check insulation for dampness.

It's often difficult to follow the path of water that penetrates a roof. However, patching the roof leak won't solve the problem. You do need to make sure the roof sheathing hasn't rotted or that other, less obvious problems in your ceiling or walls haven't developed. Once you've got a handle on the damage, it's good to detail a comprehensive plan to fix the damage, but first you need to solve the problem.

Keep the whole roof cold to avoid ice dams—Damage by ice dams can be prevented in two ways: Maintain the entire roof surface at

Three ways to vent a cathedral ceiling

Ventilation in cathedral ceilings is a little trickier than in conventional ceilings, but it is just as necessary to prevent ice damming. Careful sealing of heated spaces and good insulation and ventilation all are required.

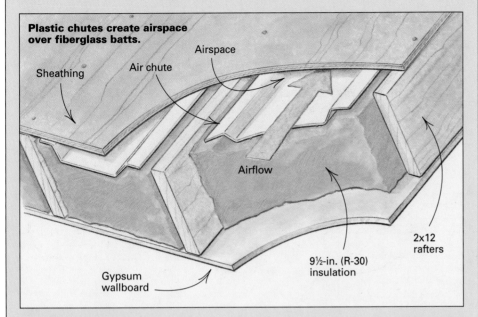

Plastic chutes create airspace over fiberglass batts.

- Sheathing
- Air chute
- Airspace
- Airflow
- Gypsum wallboard
- 9½-in. (R-30) insulation
- 2x12 rafters

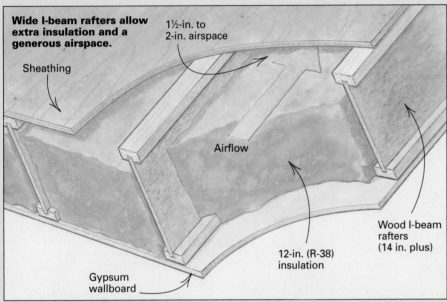

Wide I-beam rafters allow extra insulation and a generous airspace.

- Sheathing
- 1½-in. to 2-in. airspace
- Airflow
- Gypsum wallboard
- 12-in. (R-38) insulation
- Wood I-beam rafters (14 in. plus)

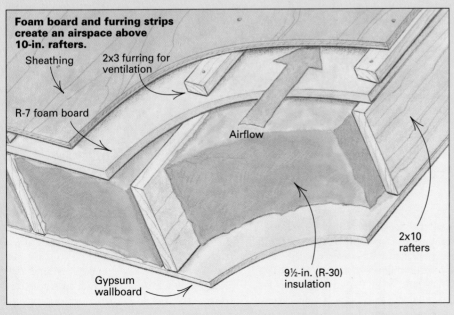

Foam board and furring strips create an airspace above 10-in. rafters.

- Sheathing
- 2x3 furring for ventilation
- R-7 foam board
- Airflow
- Gypsum wallboard
- 9½-in. (R-30) insulation
- 2x10 rafters

ambient outdoor temperatures so that an ice dam never forms; or build a roof that won't leak if an ice dam does form (bottom drawing, p. 31).

The first choice is definitely the best. Cold roofs make sense because they make the cold outside air work for you. If you keep the roof as cold as the outdoor air, you solve the problem. Look at the roofs of unheated sheds. Ice dams don't form on them because the air inside the roof is as cold as the air outside.

It's relatively easy to keep a roof cold in new construction: Design the house to include plenty of ceiling insulation and effective roof ventilation, and make sure heat doesn't escape from the house into the attic. Insulation retards the flow of heat from the heated interior to the roof surface; good ventilation keeps the roof sheathing cold; well-sealed walls and ceilings keep the heat where it belongs.

In an existing house this approach may be more difficult because often you're stuck with less than desirable conditions. This opportunity is a good point to take a closer look at the issues that will guide your strategy.

When treating symptoms is the only choice—The list of efforts to deal with ice dams is long. The problem I have with most of these solutions is that they treat the symptoms of ice damming and don't deal with the root cause, heat loss.

For instance, some people assume they can fix the problem by installing a metal roof. Metal roofs are common in snow country, so they must work, right? Well, a deeply pitched metal roof does, in a sense, thumb its nose at ice dams. Metal roofs are slippery enough to shed snow before it causes ice problems. However, metal roofs are expensive, and they do not substitute for adequate insulation.

Or you might consider using sheet-metal ice belts if you don't mind the look of a shiny 2-ft. wide metal strip strung along the edge of your roof. Ice belts are reasonable choices for some patch-and-fix jobs on existing houses. This eave-flashing system tries to do what metal roofing does, which is shed snow and ice before they cause problems. Unfortunately, it doesn't always work. Often, a secondary ice dam develops on the roof just above the top edge of the metal strip, so the problem simply moves from one part of the roof to another. Ice belts are sold in 32-in. by 36-in. pieces and come with fastening hardware for about $12 per panel.

Many people install self-sticking rubberized sheets under roof shingles wherever ponding of water against an ice dam is possible: along the eaves, around the chimneys, in valleys, around skylights and around vent stacks. The theory is that if water leaks through the shingles, the waterproof underlayment will provide a second line of defense.

These products are sold in 3-ft. by 75-ft. rolls for about $80 per roll. They adhere directly to clean roof decking. Roof shingles are nailed to the deck through the membrane, which is self-healing and seals nail penetrations automatically. Grace Ice and Water Shield (W. R. Grace Co., 62 Whittemore Ave., Cambridge, Mass. 02140; 617-

Continuous vents keep roof washed in cold air. A ridge vent draws cold air from the continuous soffit vents uniformly across the underside of the roof sheathing.

Elevated rafters make room for insulation. The builder made room for a thick blanket of insulation above the exterior wall by nailing the heels of these rafters to a toe board that sits on top of the ceiling joists. This construction method also maintains an adequate ventilation space between the sheathing and the insulation so that air can flow from the soffit vent to the ridge vent.

876-1400) and Ice and Water Barrier (Bird Roofing Products Inc., 1077 Pleasant St., Norwood, Mass. 02062; 800-247-3462) are two common brands. Installing such products is a reasonable alternative for many existing structures where real cures either are not possible or cost-effective.

Heat tape is another favorite solution to ice damming. I have never seen a zigzag arrangement of electrically heated cable solve an ice-dam problem. Electricity heats the cable, so you throw more costly energy at the problem (keep in mind that ice dams are a heat-loss problem). Heat tape is expensive to install and to use. Over time the tape makes shingles brittle and creates a fire risk, and its loose fasteners allow water to leak into the roof. Take a good look at roofs equipped with heat tape. The electric cable creates an ice dam just above it. My advice is don't waste your time or money here.

Shoveling snow and chipping ice from the edge of a roof is my least favorite of all solutions. People attack mounds of snow and roof ice with hammers, shovels, picks, snow rakes, crowbars and, new to my list, chainsaws. The theory is obvious: Where there's no snow or ice, there's no leaking water. Some people have even carved channels in the ice to let trapped water flow out.

Whatever plan you decide to follow, focus on the cause. Ice dams are created by heat lost from the house. So whenever possible, develop a strategy that includes plugging all heat leaks into unheated spaces. You can use urethane spray foam in a can, caulking, packed cellulose or weatherstripping to seal leaks made by wiring, plumbing, attic hatches, chimneys, interior partition walls and bathroom exhaust fans.

There are no excuses for ice-dam problems in new construction. But in existing houses you can improve ventilation, upgrade insulation and block as many air leaks as you can. However, cures for existing structures are often elusive and expensive, and in some cases you may have to settle for merely treating the symptoms. The payback is damage prevented.

Prevention is the key in new construction— Houses in heating climates should be equipped with ceiling insulation of at least R-38, which equals about 12 in. of fiberglass or cellulose. The ceiling insulation should be of continuous and consistent depth.

As I mentioned earlier, the biggest problem area is just above the exterior wall. Raised-heel trusses or roof-framing details that allow for R-38 above the exterior wall—while maintaining room for airflow from the eave to the ridge—should be used in new construction (photo above right). (In existing structures, where there's little space between the top plate and the underside of the roof sheathing, install R-6-per-in. insulating foam). Insulation slows conductive heat loss, but an effort must be made to block the flow of warm indoor air into the attic or roof. Even small holes allow significant volumes of warm indoor air to pass into the attic. In new construction it's best to avoid ceiling penetrations (such as recessed lights) whenever possible.

Soffit-to-ridge ventilation is the most effective way to cool roof sheathing. Power vents, turbines, roof vents and gable louvers don't work as well. Soffit and ridge vents should run continuously along the length of the house (photo above left). A baffled ridge vent is best because it exhausts attic air regardless of wind direction. Exhaust pressure created by the ridge vent sucks cold air into the attic through the soffit vents.

In cathedral ceilings it's important to provide a 2-in. space or air chute between the top of the insulation and the underside of the roof sheathing. The incoming air washes the underside of the roof sheathing with a continuous flow of cold air. The construction of cathedral ceilings requires some special consideration because the ceiling and the roof are the same structure (drawings facing page).

Sources of baffles and air chutes
ADO Products
7357 Washington Ave.
Edina, Minn. 55439
(800) 666-8191
Sells high-impact polystyrene chutes.

Guaranteed Baffle Co.
P. O. Box 510
Bend, Ore. 97709
(503) 383-0095
Sells air chutes.

Insul-Mart Inc.
156 Wickenden St.
Providence, R. I. 02903
(401) 831-0800
Sells Proper-Vent polystyrene chutes.

Insul-Tray Inc.
E. 1881 Crestview Drive
Shelton, Wash. 98584
(206) 427-5930
Manufactures water-resistant corrugated-cardboard baffles.

I want to add a caution here. Cold air flowing in through soffit vents in any kind of roof can blow loose-fill fiberglass or cellulose insulation out of the way. It will find pathways through batt or roll fiberglass. Unless the pitch of the roof prohibits it, install cardboard or polystyrene baffles in the attic space above exterior walls (bottom drawing, p. 31) to protect insulation from cold air. □

Paul Fisette is program director of Building Materials Technology & Management at the University of Massachusetts in Amherst. Photos by the author.

Energy-Saving Details

A demonstration house in Canada shows new approaches to energy-efficient, environmentally sensitive construction

Total energy cost: $800 a year. Every system in the Manitoba Advanced House was chosen according to its impact on the environment. The pine roofing is recyclable and produced with less energy than asphalt roofing. The synthetic stucco mitigates air and water infiltration, and the energy-efficient windows hold heat inside in winter while keeping it out in summer.

by Kip Park

Canada has the highest energy use per capita of any country in the world. That's not unexpected, given the country's climate. In Winnipeg, Manitoba, for example—said to have the coldest winters of any capital city outside of Siberia—temperatures can drop as low as -48°F. But in summer, temperatures can soar: The highest recorded in this prairie city of 600,000 is 108°F. That kind of climate boosts both space-heating and air-conditioning costs.

During the oil embargo of the mid-1970's, the Canadian government started the R-2000 Home Program. Houses built to R-2000 standards consumed half the energy of the typical houses being built at the time. The reduction was achieved by increasing insulation, reducing air leakage, improving window performance, using more efficient heating systems and installing a mechanical ventilation system, usually a heat-recovery ventilator (HRV).

But the R-2000 homes still used too much energy. So in 1991, Energy, Mines and Resources Canada started the Advanced House program and proposed building as many as 10 low-energy, environmentally "green" homes across the country to showcase cutting-edge and traditional technologies to chop energy bills to one-quarter of those attributed to a typical house built in 1975.

The program targeted not only space-heating and air-conditioning costs but all energy use, including embodied energy, the amount of energy it takes to produce, manufacture, distribute, install, operate and, eventually, dispose of everything that goes into a house. Advanced House techniques and technology are available from Manitoba Energy and Mines Info Center (Suite 360 Ellice Center, 1395 Ellice Ave., Winnipeg, Man. R3G 0G3; 204-945-4154).

The Advanced House had to be comfortable and convenient to live in, not forcing major changes in lifestyle. And its design had to have market appeal. There's little point in advancing technology if no one is going to buy and use it.

Energy conservation can be stylish and practical—In the fall of 1992, the Manitoba Home Builders Association (231-1120 Grant Ave., Winnipeg, Man. R3M 2A6; 204-477-5110) was the first group in Canada to construct and open an Advanced House (photo, above). According to computer simulations, it would cost about $800 a year to heat, cool, ventilate and provide domestic hot water in Manitoba's Advanced House. That's far from the $2,400 for a typical 1975 house. Both figures include about $200 a year for standby utility charges, fees to keep electrical and natural-gas lines running into the house.

Monitoring sensors and thermocouples were installed around the foundation and in wall cavities to measure moisture and temperature levels; the performance of the home was scrutinized closely through the fall of 1994.

The 1,859-sq. ft., four-bedroom Manitoba Advanced House demonstrates to contractors and tradespeople that it doesn't take a genius to conserve materials and energy. To the public, it shows that energy conservation can be stylish as well as practical and comfortable.

Reducing waste and using whiskey bottles—A major concern throughout construction was the amount of construction waste. It's been estimated that as much as 20% of materials going into landfills is construction waste. Transporting that waste to landfills is an energy cost in itself. Creating landfills eats up valuable and productive land.

"We wanted to reduce the amount of waste as much as possible," said co-designer and home-energy analyst John Hockman of Appin Associates of Winnipeg (472 Academy Road, Winnipeg, Man. R3N 0C7; 204-488-4207). So the floor plan was designed to minimize construction-material waste (by using full 4x8 sheets wherever possible, for instance), and scrap lumber was used for blocking. Suppliers were advised that all packaging materials had to be made of recycled materials or be capable of being recycled.

On the roof, pine shakes (Prairie Shake Roofing, 885 Century St., Winnipeg, Man. R3H 0M3; 204-786-0813) were used instead of asphalt shingles, which require large amounts of nonrenewable energy to produce. Pine shakes cost about 15% more than asphalt shingles, but they have a life expectancy of 30 to 50 years, after which they can be recycled.

The fill under the foundation and over the plastic drain pipes, or weeping tiles, is an indication of the imagination used. About 30% of the fill is empty whiskey bottles, smashed and pulverized on-site when tumbled in a concrete mixer with pea gravel. "We expected some problems with sharp edges of broken glass," said consulting engineer Gary Proskiw, the other designer of the house and head of Proskiw Engineering Ltd. (1666 Dublin Ave., Winnipeg, Man. R3H 0H1; 204-633-1107), which has extensive experience in residential-energy use. "The only problem we did encounter was the smell. It was like a distillery."

The ground-up bottles reduced the amount of gravel required: Gravel must be transported, cleaned and sifted, all of which consume energy. So Canada's energy use was reduced slightly by whiskey drinkers.

Split-stud walls add insulation space— Predictably, there are high levels of insulation in the Manitoba house: R-60 in the ceiling, R-42 on poured-concrete basement walls and rigid insulation under the entire basement-floor slab, totaling R-10. The 12-in. thick exterior walls have an R-value of 46. The walls are a split-stud system (photo, bottom right, where the exterior 2x4s are separated from the interior 2x3s by 6-in. long, ⅛-in. dia. metal rods. Manufactured by Ten Lives Industries (60 Heaton Ave., Winnipeg, Man. R3B 3E3; 204-956-2860), these wall trusses are easier to install than a site-built double wall. The wall trusses use less lumber than a double wall, yet the trusses are stiffer because of the metal rods, which conduct less heat through walls than a full stud. In conventional wall systems, about 18% of the wall area is made up of studs; a 2x6 stud has an insulating value of about R-7.

The high-performance thermal envelope features rim joists covered with housewrap (photo, bottom left), which adheres to the structure with acoustical sealant. On the interior side, a 6-mil polyethylene air/vapor barrier is sealed to the housewrap (drawing, right). The barrier is nearly hole-free; it goes up the wall and into the joist cavities, where it's stapled to the joist faces only.

The combination of housewrap and air/vapor barrier is engineered for a cold climate. The poly stops moisture from entering the wall, and the housewrap stops air from flowing in or out, yet it allows any moisture in the wall to escape.

The air/vapor barrier is sealed to the electrical boxes, which have adhesive gaskets (Nu-Tek Plastics Inc., 25-11151 Horseshoe Way, Richmond, B. C. V7A 4S5; 604-272-5550). The Manitoba Advanced House, when tested with a blower door, had an air-change rate (ach50) of 0.78 per hour at 50 Pascals; requirements call for 1.5 ach50 (the tightest house in Canada, also in Winnipeg, had less than 0.20 ach50 when it was built in 1982).

Conserving natural resources—Very little large-dimension lumber was used in the Manitoba Advanced House. "The largest pieces

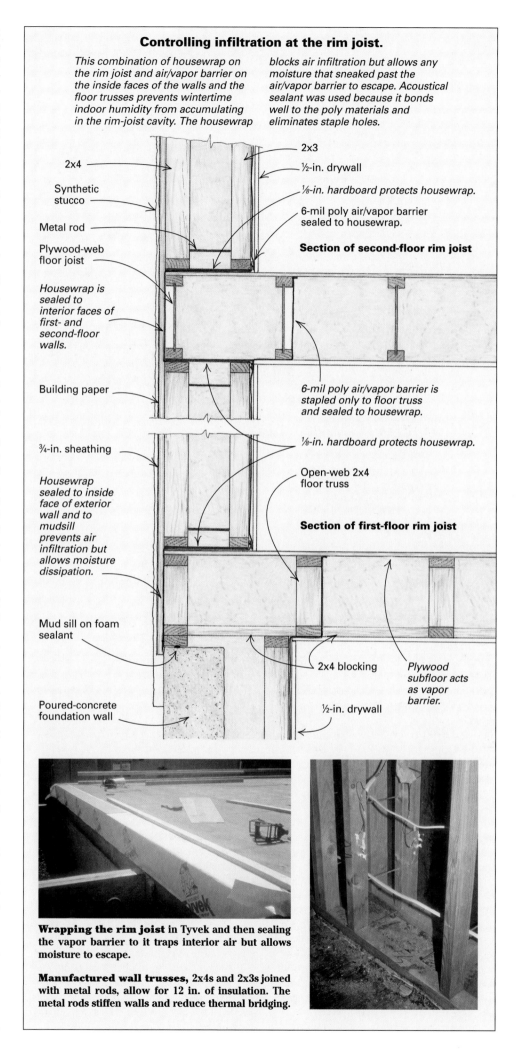

Controlling infiltration at the rim joist.

This combination of housewrap on the rim joist and air/vapor barrier on the inside faces of the walls and the floor trusses prevents wintertime indoor humidity from accumulating in the rim-joist cavity. The housewrap blocks air infiltration but allows any moisture that sneaked past the air/vapor barrier to escape. Acoustical sealant was used because it bonds well to the poly materials and eliminates staple holes.

2x4

Synthetic stucco

Metal rod

Plywood-web floor joist

Housewrap is sealed to interior faces of first- and second-floor walls.

Building paper

¾-in. sheathing

Housewrap sealed to inside face of exterior wall and to mudsill prevents air infiltration but allows moisture dissipation.

Mud sill on foam sealant

Poured-concrete foundation wall

2x3

½-in. drywall

⅛-in. hardboard protects housewrap.

6-mil poly air/vapor barrier sealed to housewrap.

Section of second-floor rim joist

6-mil poly air/vapor barrier is stapled only to floor truss and sealed to housewrap.

⅛-in. hardboard protects housewrap.

Open-web 2x4 floor truss

Section of first-floor rim joist

2x4 blocking

Plywood subfloor acts as vapor barrier.

½-in. drywall

Wrapping the rim joist in Tyvek and then sealing the vapor barrier to it traps interior air but allows moisture to escape.

Manufactured wall trusses, 2x4s and 2x3s joined with metal rods, allow for 12 in. of insulation. The metal rods stiffen walls and reduce thermal bridging.

Photos this page: John Hockman and Gary Proskiw

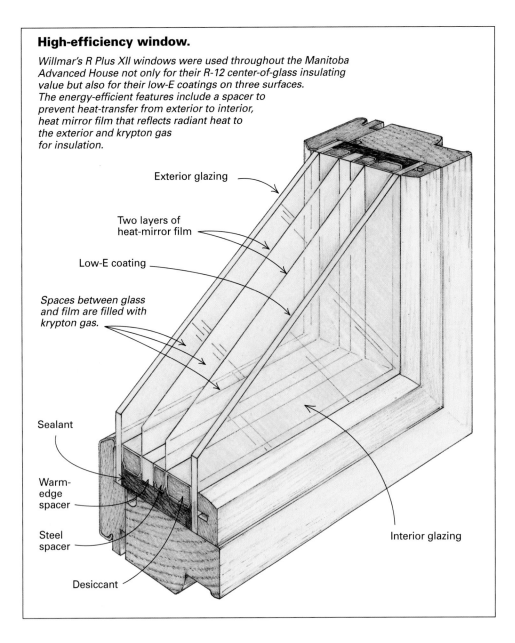

High-efficiency window.

Willmar's R Plus XII windows were used throughout the Manitoba Advanced House not only for their R-12 center-of-glass insulating value but also for their low-E coatings on three surfaces. The energy-efficient features include a spacer to prevent heat-transfer from exterior to interior, heat mirror film that reflects radiant heat to the exterior and krypton gas for insulation.

Exterior glazing

Two layers of heat-mirror film

Low-E coating

Spaces between glass and film are filled with krypton gas.

Sealant

Warm-edge spacer

Steel spacer

Desiccant

Interior glazing

---

of lumber are 8-ft. long 2x8s," said Glenn Buchko, executive director of the Manitoba Home Builders Association, which acted as general contractor for the project. The 2x8s were in the garage-door headers.

The first-floor joists are open-web trusses made of 2x4s; the second-floor joists have finger-jointed cords and a plywood web. Why use two different trusses? The open-web joists allowed for large ducts and conduits; the closed-web joists use less material. Both types of truss "dramatically reduced our need for large-dimension lumber, thus lowering the demands we impose on our forests," said Buchko.

Water conservation is also a major factor in the Manitoba Advanced House. Water-efficient toilets, shower heads and faucets reduce water usage without sacrificing performance. Low-flush toilets use about 1.5 U. S. gallons per flush, compared with 3.5 gallons for conventional toilets.

Infrared sensors (Crane Canada Inc., 5850 Cote-de-Liesse Road, Montreal, Quebec H4T 1B2; 514-735-3592) on bathroom water faucets automatically turn water on and off when hands are removed from beneath the faucet.

Landscaping includes drought-resistant native plants, which are adapted to Manitoba's dry summers. These plants reduce watering requirements and provide summertime shade and a comfortable, attractive climate around the house.

There's even a system to collect and store rainwater from the roof and from the sump pit for use in watering the lawn and garden.

Windows for all seasons—The Manitoba Advanced House is oriented to maximize passive-solar heat gain—but with a difference. The program called for window treatments that would control summertime overheating. The most energy-efficient windows in Canada, Willmar's R Plus XII (drawing, above), were installed throughout (Willmar Windows, 485 Watt St., Winnipeg, Man. R2L 2A5; 204-668-8230). They're considered quadruple-glazed, though two of those glazings are actually Heat Mirror 88 film, which reflects radiant heat while it transmits 88% of the light, suspended between two panes of glass. The inner light has an additional soft low-E coating, and the air spaces are filled with krypton gas. A special spacer-bar system was developed to keep the edges of the lights warmer than conventional-window edges, and

center-of-glass insulation values hit R-11.5. R Plus XII windows cost about one-third more than conventional double-glazed windows.

To prevent solar overheating in summer, all windows are equipped with removable solar screens, which block about 30% of solar radiation. That reduces the cooling load by more than one-third and, when combined with the house's ventilation system, also eliminates the need for air conditioning.

Heating and ventilating—On the mechanical side, a 94% efficient natural-gas water heater with a custom-built air handler (Mor-Flo Industries Inc., division of American Water Heater Group, P. O. Box 4056, Johnson City, Tenn. 37602; 615-283-8000) heats both the home (forced hot air) and the domestic hot water. The unit's high efficiency is achieved in part by circulating warm greywater, the wastewater from such things as washing machines and dishwashers, through a heat exchanger (drawing, facing page) to preheat water going to the boiler.

Indoor-air quality is maintained by a heat-recovery ventilator (HRV), which preheats incoming air. The modified Lifebreath 195 HRV (Nutech Energy Systems Inc., 511 McCormick Blvd., London, Ont. N5W 4C8; 519-457-1904) mixes fresh air with return air, and all air is then filtered through an 85% efficient bag filter to remove dust and fine particulate matter. The air then passes through an activated-charcoal filter to remove odors and chemicals before it is heated and distributed.

The home is divided into four heating zones—all forced hot air except for hot-water baseboard in the fourth bedroom, which I'll explain in a minute—allowing one area of the home to receive fresh, heated air, while another (unoccupied) area still can receive fresh, unheated air. To ensure high air quality and to prevent the chronic problem of mold growth in cold closets, return-air vents, which pull stale air from the room, are installed in all bedroom closets. The location of these vents improves overall air quality, too. "Clothing can be a significant source of indoor-air pollution because of chemicals used in dry cleaning," Proskiw noted. "Here, we're eliminating them at the source."

The fourth bedroom, on the main floor, can be used as a den, a hobby room or a smoking room. It has its own separate exhaust-air system, so any noxious odors or chemicals produced by hobbies or smoking are exhausted directly outside. For the same reason, this is the room heated with hot-water baseboards.

Great care was taken in selecting materials used inside the home to ensure good indoor-air quality. For instance, the technical requirements as issued by Energy, Mines and Resources required that products containing urea-formaldehyde-based resin glues, such as the chipboard used in the kitchen cabinets, be sealed to limit formaldehyde outgassing. Indoor air could not contain any more than 0.05 parts per million of formaldehyde.

Using less electricity—"There are two major factors affecting the energy performance of any

Drawings: Christopher Clapp

house—the structure itself and the people who live in it," said Hockman. In the Manitoba Advanced House, a Power Sentry power-usage indicator was installed above a desk in the kitchen to show present and cumulative use of electricity (Northwest Extension Inc., 15 Central Way #201, Kirkland, Wash. 98033; 206-828-9190). The power-usage indicator also presents data in dollars and cents.

In other installations, this type of information system, which makes homeowners aware of just how much electricity is being used at any moment, has resulted in an immediate 10% to 20% reduction in consumption. "It's an awareness thing," said Proskiw.

The appliances by General Electric are among the most energy-efficient models on the market. The clothes washer is rated at 74kwh per month, compared with conventional washers that can use 150kwh. The refrigerator is rated at 55kwh compared with similarly sized conventional units that can use as much as 85kwh per month.

Artificial lighting was considered important because Winnipeg receives only about eight hours of daylight per day during December. High-efficiency fluorescent and halogen lighting fixtures are used throughout most of the house. Less-efficient incandescent lighting is restricted to fixtures that are seldom used, such as closet lights, because the incandescent lights are a lot less expensive. The Manitoba Advanced House also makes extensive use of dimmers and automatic light switches to promote energy savings and to increase convenience.

In Winnipeg's cold winter climate, engine-block heaters and and vehicle-interior warmers keep cars from freezing up. Because these comfort devices use lots of electricity, an indoor timer automatically switches on the heaters about two hours before vehicles will be used.

Future repercussions—There are a multitude of features in the Manitoba Advanced House, many of them simple in concept. "All the features had to be as practical as possible, with the widest potential application," said Hockman. "This type of house will be aimed at the mainstream in the very near future."

The principles learned and demonstrated in the Manitoba Advanced House eventually could be transferred to the renovation industry. Because much of the technology incorporated in the Manitoba Advanced House is cutting-edge, it is, therefore, expensive. The lessons learned in Winnipeg, however, will help that technology develop further, and the cost of renovating existing homes will drop.

Although its monitoring systems made the Manitoba Advanced House about 25% more expensive than a conventionally constructed home, the house is truly a demonstration that energy-efficient, environmentally friendly, "green" houses are practical and ready to stand the test of the marketplace. ☐

Kip Park lives in Winnipeg, Manitoba, and he writes about housing, construction and energy-technology issues. Photo by the author except where noted.

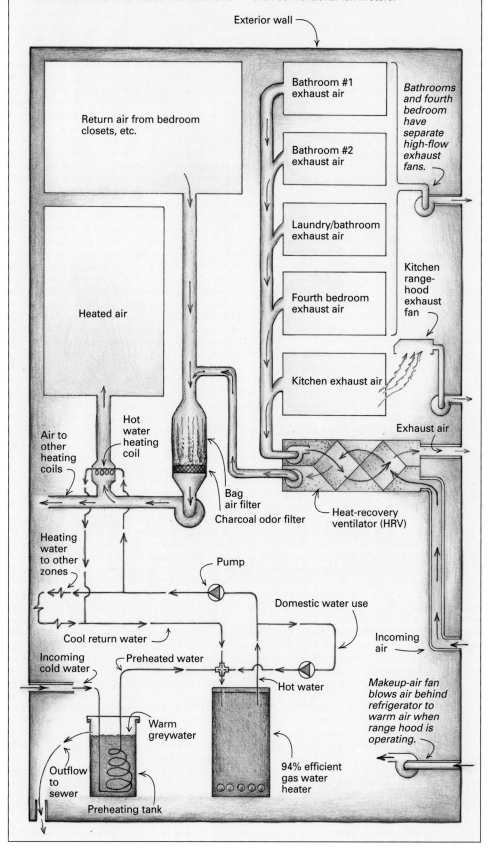

An energy-efficient heating and ventilating system.

Using one-third the energy of a typical house built in 1975, the Manitoba Advanced House incorporates heat-capturing devices, energy-efficient appliances and multiple heating zones. Incoming water is warmed in a custom-built preheater that captures heat from dishwasher and laundry wastewater; the preheated water then goes to a highly efficient gas water heater and is distributed to the domestic hot water and the heating system. The ventilating system, required in the airtight Manitoba house, pulls outdoor air through a heat-recovery ventilator, which warms the air with the residual heat of stale air that's exhausting from the house. The warmed incoming air mixes with recirculated indoor air and is filtered twice, then sent to one of four separate heating/ventilating zones. All fan motors are more efficient than conventional fan motors.

Exterior wall

Return air from bedroom closets, etc.

Heated air

Bathroom #1 exhaust air

Bathroom #2 exhaust air

Laundry/bathroom exhaust air

Fourth bedroom exhaust air

Kitchen exhaust air

Bathrooms and fourth bedroom have separate high-flow exhaust fans.

Kitchen range-hood exhaust fan

Exhaust air

Air to other heating coils

Hot water heating coil

Bag air filter

Charcoal odor filter

Heat-recovery ventilator (HRV)

Heating water to other zones

Pump

Domestic water use

Cool return water

Incoming cold water

Preheated water

Hot water

Incoming air

Makeup-air fan blows air behind refrigerator to warm air when range hood is operating.

Warm greywater

Outflow to sewer

Preheating tank

94% efficient gas water heater

A Rumford Fireplace, Southwestern Style

A California mason combines the adobe look of a kiva
with the heating efficiency of a Rumford

by Vladimir Popovac

Photo this page: Roe A. Osborn

The city inspector said it was the first house he'd ever seen built entirely from faxes. The supervising contractor seemed to be on the phone every other minute as the owners and the architect became absorbed in the project. For the crew, frustrations occasionally outpaced pride of accomplishment. Carpenters immersed in a complex job often had to revise the previous day's work as fresh ideas poured in.

My partner, Marshall Dunn, and I were luckier. The architect and the contractor always managed to keep changes one step ahead of the masonry work. We had to change our plans several times, shift our setup lines and recalculate our numbers, but we never had to tear out any work. We were asked to build a fireplace that would fit in with the Southwestern style of the house (photo facing page). The rounded shape of the fireplace, its arched opening and its plastered finish would be reminiscent of a Native American ceremonial kiva structure.

The circular shapes found in the architecture of Native Americans reflect their belief in the cyclical characteristics of life and nature, such as the progression of the seasons. Ironically, as the fireplace itself progressed from drawing board to finished fireplace, its appearance changed dramatically. But despite the changes, we were determined to build a fireplace that would burn efficiently and cleanly.

The only fireplace design that would give us the performance we were seeking is the Rumford, an extremely efficient fireplace design developed two centuries ago (sidebar right). So I started looking into the possibility of building this little kiva fireplace with the proportions and configuration of a classic Rumford.

Start with a good foundation—Every good fireplace begins with a good foundation, and our kiva was built on top of an 8-in. thick reinforced-concrete slab that was tied in to a fully grouted (all voids filled with mortar) and reinforced 8-in. concrete-block perimeter (drawing p. 40).

We formed the slab on top of a ½-in. thick cementitious backerboard, which we left in place after the slab had cured. The space directly under the fireplace floor is sometimes left open as a place to pile up ashes that get swept through an ash drop in the firebox floor. However, with a living area below the fireplace, we opted to leave the space under the slab empty.

We set our slab about 17 in. off the subfloor, which was 2 in. lower than the finished height of the stone hearth and bench that would run in front of the kiva and along one wall. The hearth and bench sit on top of a mortar bed that is supported by backerboard on top of a short framed wall that forms the curve of the bench.

We built the firebox floor out of 1¼-in. thick firebrick pavers set into a thin layer of refractory cement. The firebox floor would then end up ¾ in. lower than the stone hearth, a margin that helps to keep the ashes in the fireplace.

Before starting the firebox floor, we built the outer walls of the fireplace with fully grouted reinforced block. We normally use 8-in. wide block for the walls, but the limited space on this job permitted only 4-in. block. Unlike 8-in. block, 4-in. block is difficult to fill in one pour after the walls are built, so we grouted the block course by course as we went up.

Adapting the Rumford to a kiva—Once the firebox floor was set, we were ready to lay out our lines for the firebox walls. Rumford fireplaces generally run from 24 in. to 48 in. wide, and manufactured parts are available from Superior Clay Corporation (Uhrichsville, OH; 800-848-6166) that fit this size range and facilitate construction. These parts consist of a one-piece throat with the distinctive rounded breast, the damper, the smoke chamber and the matching flue liners. Everything is precast, solidly built and fit together with sublime simplicity.

This kiva, which was designed by the architect, called for an opening only 16 in. wide. With the smallest manufactured parts made for a 24-in. opening, at first we figured we'd have to make our own Rumfordlike parts. Besides, the base (or footprint) of a standard Rumford firebox is trapezoidal, looking sort of like a squashed rectangle, wide in front and narrow in back. The traditional kiva firebox structure is supposed to be semicircular in keeping with basic Native American architecture.

We came up with a way to honor that principle by building our firebox in the shape of a half-circle while still using precast parts. We built the firebox of firebricks cut in half and installed on edge with their ends showing to achieve the required radius. The firebricks are about 9 in. long and can be cut in half with a saw, but we found it quicker and easier to break the bricks in half by tapping them with a mason's hammer.

The half-brick length of 4½ in. on each side of the 16-in. specified opening gave us an outside perimeter of 25 in. for the firebox, which was enough to accommodate the smallest precast Rumford throat with no problem (bottom photo, p. 40). We had all the precast parts on hand and set the throat piece on the lines we drew on the firebox floor to make sure it would fit before we started laying the firebrick.

Refractory cement is best for high heat— For the firebox as well as for all masonry joints that will be exposed to fire and high heat, we always use refractory cement. Refractory ce-

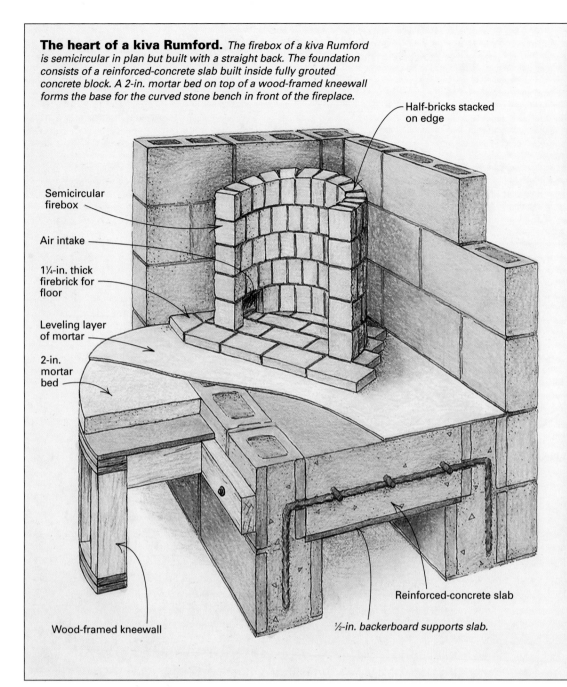

The heart of a kiva Rumford. *The firebox of a kiva Rumford is semicircular in plan but built with a straight back. The foundation consists of a reinforced-concrete slab built inside fully grouted concrete block. A 2-in. mortar bed on top of a wood-framed kneewall forms the base for the curved stone bench in front of the fireplace.*

Half-bricks stacked on edge

Semicircular firebox

Air intake

1¼-in. thick firebrick for floor

Leveling layer of mortar

2-in. mortar bed

Wood-framed kneewall

Reinforced-concrete slab

½-in. backerboard supports slab.

A plywood form for the arched opening. To make the rounded opening, firebricks were set in refractory cement around a form of two sheets of plywood separated by a 2x spacer.

A precast throat sits on top of the firebox. A manufactured Rumford throat with its characteristic curved front, or breast, is a key ingredient for turning this kiva fireplace into an effective, clean-burning heater.

ment comes in a container ready to spread. Unlike most cement and mortar that use water as part of a chemical reaction to turn the compound solid, refractory cement becomes solid solely through evaporation. Refractory cement, available at masonry-supply stores, is resistant to extreme changes in temperature, and the cement we used stays solid up to 2000°F.

Refractory cement is supposed to be used thin, so we made the mortar joints no more than ⅛ in. wide. Because we were stacking rectangular brick along a circular line, the joints between the firebrick filled with refractory cement widen in a wedge shape as the bricks fan out. We parged the back of the firebrick and the wedge-

shaped joints with a mixture of portland cement and fire-clay mortar because we wanted to avoid fat joints of refractory cement and because that area would not be exposed to flames or high temperatures. The space between the back of the firebox wall and the outer block wall was filled with loose rubble and mortar, which gives the firebox room for expansion.

We notched an opening for an air intake into the second course of brick on the firebox walls. The air intake that we used was roughly the same size as a brick, about 2 in. high, 4 in. wide and 9 in. long. It works like a drawer that pulls out to allow fresh outside air into the firebox at the level of the firebox floor. We recessed the

air intake in the firebox wall so that when closed, it is barely visible.

On the backside of the firebox, we covered the mechanism for the air intake with a curved piece of sheet metal to keep rubble and mortar away from it. We also threaded some long bolts into the end of the intake to anchor it securely in the rubble. An opening was left at the proper location in the outer fireplace wall for a 2-in. pipe to the outside, which was screened to keep out small animals and large insects.

The precast throat gets its corners rounded off—After completing the firebox walls, we bedded the precast throat in place with refrac-

Drawings: Christopher Clapp

Curve meets trapezoid. As you look down the trapezoidal precast throat, the top of the curved firebox is visible. Cut firebrick set in refractory cement smoothed out this transition.

A damper covers the top of the throat. A manufactured damper is made to go over the opening of the throat. The floor of the smoke chamber surrounding the damper is finished in refractory cement.

tory cement (bottom photo, facing page). Cast to fit the standard Rumford shape, the bottom of the throat is a trapezoid. The inside corners of the throat installed on top of the semicircular kiva firebox left triangular spaces on top of the flat firebox wall (photo top left) that had to be filled in to maintain the smooth interior surface important to the Rumford design. We filled these spaces with scraps of firebrick that were cut and set in refractory cement to achieve a uniform transition between the firebox and the throat.

Next, we built the arch for the firebox opening. We made a half-oval form the same height as the firebox out of two plywood scraps separated by a spacer and then stacked firebrick around it (top photo, facing page). Where the backside of the arch met the firebox at a 90° angle, we cut scraps of firebrick and set them in refractory cement to soften the corners. We also rounded the front edge of the bricks in the arch with an electric grinder (photo right).

Kiva shape has a brick foundation—Next, we scribed lines on the walls and the floor on both sides of the fireplace to form the kiva shape that the architect was seeking. The lines had been altered two or three times during the previous week. But once contours of the kiva were laid out, we started building with confidence. We stacked bricks in a bellied curve that joined the sides of the fireplace arch to the lines on the walls. On one wall the line was stepped to make way for three tapered shelves that would each be topped with stone to match the bench. Each brick on each course was eyeballed in place. Because they would be covered with two coats of plaster, they just had to be close, not perfect.

We filled the spaces between the brick front, the firebox and the outer walls with rubble and mortar as we went up. When we reached the top of the precast throat, we brought the whole assembly up level and flat. The next step was positioning the damper (photo top right) and the precast smoke chamber.

The smoke chamber is cut down to size— Normally, the damper is just mortared snug and centered on top of the precast throat. Then the two-piece smoke chamber is set around it, centered and plumb. But for this fireplace, space was at a premium, so we had to set the smoke chamber slightly askew, slanting toward the corner of the room. We also narrowed the width of the smoke chamber to make it fit.

We calculated that the smoke chamber could be narrowed by 1½ in. and still have plenty room for the damper, so we ripped that amount off the centerline of one side of the smoke chamber front and back. Like the rest of the precast parts, the smoke chamber is made of high-fired red clay that resembles terra-cotta. It is easily cut with a diamond blade mounted on a circular saw. A dry fit of the damper and the slimmed-down smoke chamber let us know that the damper could operate without interference. Then we mortared everything in place, again using refractory mortar for all joints that would be exposed to direct heat (bottom photo, p. 42).

A metal flue saves space—We went back to stacking the brick front and filling the space between it and the block wall with rubble and mortar. This time, we left a recessed area to form

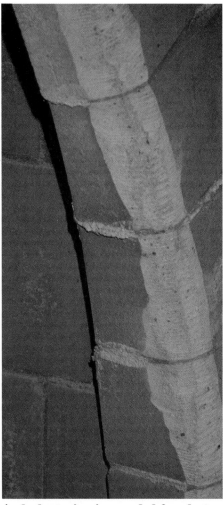

Arched opening is rounded for plaster. The edges of the firebrick around the opening of the firebox were rounded with a grinder to make a softer corner when the fireplace is eventually plastered.

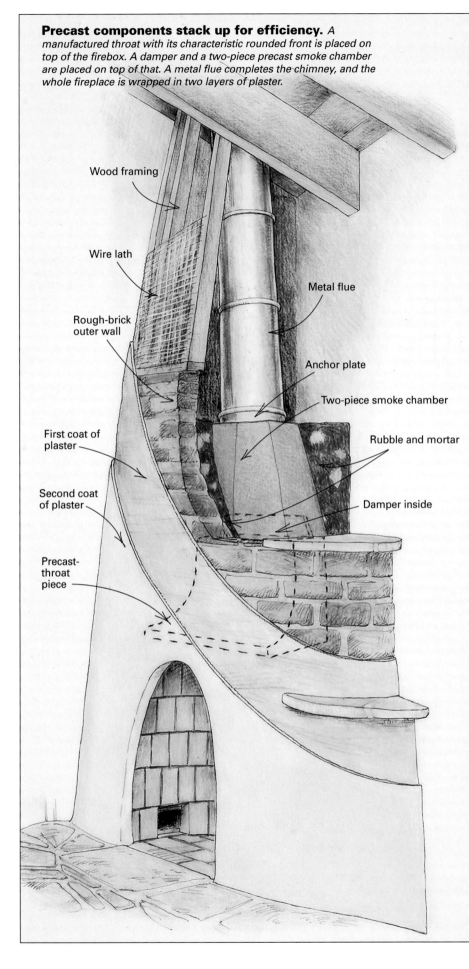

Precast components stack up for efficiency. *A manufactured throat with its characteristic rounded front is placed on top of the firebox. A damper and a two-piece precast smoke chamber are placed on top of that. A metal flue completes the chimney, and the whole fireplace is wrapped in two layers of plaster.*

Wood framing

Wire lath

Rough-brick outer wall

Metal flue

Anchor plate

Two-piece smoke chamber

First coat of plaster

Rubble and mortar

Second coat of plaster

Damper inside

Precast-throat piece

Wood framing tops the masonry. The masonry was stopped short of the ceiling so that the contractor could insert the metal flue for the chimney. The wood frame was wire-lathed and then plastered with the rest of the fireplace.

Anchoring the anchor plate. Wire that has been twisted through the corners of the metal-flue anchor plate extends into the rubble and mortar around the smoke chamber to hold the plate in place.

Refractory cement for high-heat areas. For areas of the fireplace that will be exposed to high heat, such as the joint between the two halves of the smoke chamber pictured here, refractory cement that comes premixed in a container holds up better than ordinary mortar.

a decorative niche centered over the fireplace opening. When we reached the top of the smoke chamber, we positioned an anchor plate for attaching an 8-in. inside dia. Metalbestos flue (Selkirk Metalbestos, Dallas, TX; 800-848-2149) (center photo, facing page). A metal flue had been specified instead of a masonry chimney to reduce weight and to accommodate the slant and the configuration of the chimney. Outside the house, the chimney was to be wood-framed, wire-lathed and stuccoed.

I figured that this fireplace, with its smooth Rumford lines, would have drawn well with a 6-in. flue, but we used the standard formula that the cross-sectional area of the flue should be one-eighth of the area of the fireplace opening. The height from the top of the firebox to the top of the chimney was just under 14 ft.

We set the anchor plate in mortar and secured it by twisting six strands of tie wire into each of four screw holes in its base plate. The wire was buried deep in the surrounding rubble mortar. We softened the corners of the smoke chamber with pieces of firebrick and refractory cement for a smooth transition to the anchor plate and flue. We continued our masonry to within a couple of feet of the ceiling to give the contractor room to set the flue and to frame the top of the kiva (top photo, facing page).

Stone benches are installed before plastering—With the contractor responsible for finishing the chimney, we turned our attention to the hearth and bench. The stone that the owners chose for these details came from the Middle East. The distributor called it "biblical stone." Each piece sported a shiny diamond-shaped gold label proclaiming that it was at least 300 years old. I'm sure that the stone, which looked to me like partially metamorphosed limestone, is actually over 3 million years old. I suppose the label meant that the stone had been part of some edifice that was at least 300 years old.

We mortared in the stone for the bench and for the shelves before the plastering commenced so that the plaster would wrap over the top of the stone where the two meet. The plasterers who did all the plastering and stucco work inside the house (Sunset Plastering, San Francisco, CA; 415-731-9049) did the kiva as well.

Before applying plaster, our plasterers brushed on a coat of Weld-Crete (Larsen Products, Jessup, MD; 800-633-6668) to ensure a solid bond. They followed the Weld-Crete with two coats of gypsum plaster. The finish has a slightly irregular, hand-tooled look (photo right).

The purity of the plastered kiva almost persuaded the owners not to use any color on the fireplace, but the work of painters Vicki Beggs and Suzy Papanikolas of San Francisco proved irresistible. Once the owners saw their work in

Rumford gets plastered. Two coats of plaster finish off the outside of the fireplace and create the distinctive kiva look. The stone shelves and benches were installed beforehand and masked off so that the plaster would wrap over the stone.

other parts of the house, the kiva had to be painted as well.

These talented women were equipped with an array of vessels, cans, jars, tubes of colors, powders, tints, brushes, sponges and rags, but the main ingredient they used was their experience. The fireplace is finished with an umber-tinted base topped by an ochre wash. The colors are so subtly applied and blended that they glow even without the benefit of firelight. □

Vladimir Popovac, a mason, works with his partner, Marshall Dunn, in Senora, California. Photos by the author, except where noted.

Sunspace House

Roll up the door to the solarium, and the center of this inexpensive, energy-efficient house becomes an outdoor living room

by David Hall

As much as any factor, the climate of a place influences the way the houses look. Here in the Pacific Northwest, for example, we pay particular attention to the roofs. Builders and architects have designed many a roof and awning to shelter passageways and sidewalks, and wide eaves to keep the frequent rains off the sides of the houses. And because this is a timber-rich landscape, the roofs usually are constructed out of wood. But the wide eaves (and the resulting deep rafter tails) end up stealing the light from the interior of the house.

As a result, daylight is a precious commodity in our houses. When I began designing a new home for Roger and Karen Adams, keeping the rain off the house while ensuring plentiful natural light inside was one of my priorities.

The bucks stop at $100,000—Providing plenty of daylight wasn't the only challenge. Karen and Roger had a $100,000 construction budget, period. Fortunately, we had our first design meeting before they had bought a homesite. The parcel they were considering had several obstacles, including sloping terrain and entangling ease-

Luminous eaves. The Pacific Northwestern climate can be counted on to deliver a steady diet of clouds and rain for much of the year. The translucent fiberglass eaves that border the roof keep rain from the house without plunging rooms into darkness. Purlins carry the weight of the eaves. Photo taken at A on floor plan.

ments, which would have caused an increase in the price of the house.

Roger and Karen passed on the first site, eventually settling on a long, narrow four-acre parcel on Camano Island in Washington's Puget Sound. The site is approached from the southwest, through a meadow that abuts a stand of second-growth conifers and hardwoods. To promote economy and to get as much sunlight as possible, we decided to place the house in the meadow at the forest's edge (photo facing page). The meadow occupies the flattest portion of the parcel, and it's also fairly close to the road. This kept the sitework to a minimum and the driveway within reason.

To meet a tight budget, the building's footprint and its assembly details had to be simple. To this end, I devised a rectangular floor plan of 1,150 sq. ft. I stretched it out, orienting the long side of the 20-ft. by 56-ft. plan toward the southwest (floor plan below). The layout follows a 4-ft. sq. grid. This modular theme was repeated in the floors, walls, ceilings and windows.

The plan centers around the sunspace, which lets in daylight through corrugated fiberglass roof

SPECS

Bedrooms: 1
Bathrooms: 1½
Heating system: Passive solar with propane-fired radiant heat and woodstove backup
Size: 1,150 sq. ft.
Cost: $87 per sq. ft.
Completed: 1993
Location: Camano Island, Washington

Photos taken at lettered positions.

North

0 2 4   8 ft.

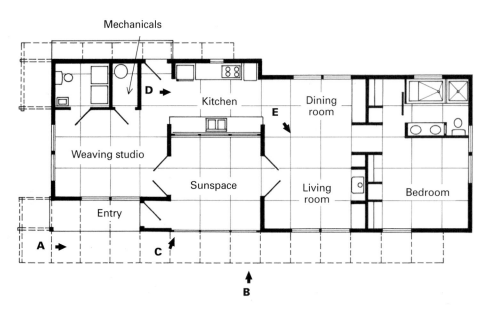

The sunspace touches most of the rooms. *Interior windows and French doors in rooms that embrace the sunroom let daylight in the house.*

Photo: David Hall. Drawing: Jeff Bellantuono

Open to the meadow. The segmented window wall of the sunspace can be retracted, creating a quasi-front porch covered with translucent corrugated fiberglass. Corrugated metal roofing continues the texture across the rest of the house. Photo taken at B on floor plan.

panels (photo left). On clear winter days the room heats up enough just from the sun to make it a comfortable place to sit and read. The space can be opened to the outdoors by lifting the handle on the glazed overhead garage door. Open to the meadow on a warm day, the sunspace becomes the modern equivalent of the traditional front porch.

Flanking the sunspace are the living room and Karen's weaving studio. Both rooms borrow light and space from the sunroom through double French doors. The kitchen shares the third interior wall of the sunroom. A 6-ft. by 12-ft. fixed window, made from a glazed garage door, allows light from the sunroom into the kitchen (photo below). Not only does the glazed garage door carry on the look of the operable door on the exterior wall, but it also made sense economically. At $625 for the cost of the three glazed sections, the window came to $8.70 per sq. ft. A comparable custom window would have cost Karen and Roger at least $1,500.

The galley-style kitchen serves as a passageway to the back door and weaving studio. We used open shelves and drawers without traditional doors and fronts for economy and efficiency.

The living room and dining room are essentially one space with shared views of meadow and

The sunspace collects and distributes the sun's warmth. An insulated concrete slab, tinted rusty red, serves as the finished floor throughout the house. Circular, wall-mounted fans near the sunspace's ceiling convey heated air to the studio and living room by way of ducts concealed in the walls. The windows at the top can be opened to keep the room from overheating. Photo taken at C on floor plan.

The kitchen is in the middle. A galley-style kitchen overlooks the sunspace, which lets abundant light into the rest of the house. The glass wall that separates the two is composed of three sections of a segmented, glazed garage door. Traditional cabinet doors and drawer fronts were omitted purposefully to emphasize the utilitarian nature of the house. Photo taken at D on floor plan.

forest. These rooms are open to the kitchen, but they are separated from the bedroom and bath by cabinet partitions that serve as closet space on the bedroom and bath side. Concrete floors and sloped ceilings continue uninterrupted from room to room.

Heating the house—Because it acts as a buffer against the weather, the sunroom contributes to the overall energy efficiency of the house by providing protection from the harsh changes in temperature and wind. Thermostatically controlled fans near the sunspace ceiling direct the hot air that collects in the sunspace to the weaving studio and the living room by way of ductwork hidden in the walls.

The sunspace roof continues past the ridge, which makes a space for clerestory windows that let in light. Perhaps more importantly, the clerestory windows can be opened to vent the sunspace when it gets too hot to be comfortable.

In addition to the heated air from the sunspace, the solar gain through the south-facing windows falling on the insulated concrete floor further tempers the heated portions of the house. Between the solar gain and the hot air from the sunspace, the house requires no backup heating for all but a few months of the year. When it does

get cold, Roger and Karen can fire up the baseboard heaters or stoke the woodstove (photo below). The stove's location in the living room provides even heat throughout the house. And with an almost unlimited supply of alder in their backyard forest, Roger and Karen have yet to rely on the baseboard heat.

Translucent eaves, plywood walls—The eaves extend several feet beyond the walls of the house, but instead of being made of shingles, sheathing and rafters, these eaves are a lattice of galvanized steel tubes and 2x4 purlins covered with translucent fiberglass panels (photo p. 44). These fiberglass panels tuck under the corrugated-metal roofing. The overall appearance is light and airy. And by providing a galvanized steel frame for the translucent eaves, the problem of exposed wooden rafters rotting in our wet climate was eliminated.

We used 4x8 sheets of MDO plywood, which has a smooth, resin-impregnated skin, as both the sheathing and the siding on the house. Satin exterior-grade varnish protects the plywood, and cedar battens cover the vertical seams between the sheets.

Concrete floors are an opportunity to use the structure as the finish, which can keep overall

costs down. We dressed up the floors by adding a red oxide pigment to the concrete prior to placing the slab. Originally we'd imagined a smooth, hand-troweled finish. But the concrete didn't behave as predicted. A stubborn cold snap kept the concrete from setting up even though the workers tried troweling it after dark, illuminated by truck headlights. Builders Gerry and John Robbins weren't satisfied with the finished surface, so they used floor grinders to level out the slab, which revealed the colors of both the aggregate and the pigment. The result, finished with a clear acrylic sealer, is better than we'd originally hoped for.

For Roger and Karen, living the simple life in a simple house has required some adjustments. With limited living and storage space, hard decisions concerning furniture, books and other items collected over the years had to be made. A large garage sale helped to trim their possessions. But when funds permit, this homestead will start to grow again. Karen and Roger's additional plans call for a detached carport with storage space and a shop. ☐

Architect David Hall is a partner in the Henry Klein Partnership in Mount Vernon, Washington. Photos by Charles Miller except where noted.

Bringing down the ridge beam. A 4x12 ridge beam supports the rafters, but instead of concealing it in the roof, the architect lowered the beam to reveal the structure. The weight of the roof is conveyed to the beam by 2x4 posts, which are sandwiched by triangular particleboard gussets. Here in the living room, a woodstove provides backup heat in the winter. Photo taken at E on floor plan.

A Primer on Heating Systems

You'll save money and stay more comfortable by matching the heating system to your climate, your house and your needs

by Alex Wilson

I learned to appreciate good heating systems at an early age—by growing up in a 1710 Pennsylvania log home that lacked one. I remember huddling around our oil-fired space heater on winter nights to soak up as much warmth as possible before dashing upstairs to the small corner room my brother and I shared. I even have a distinct—though no doubt exaggerated—memory of a particularly frigid winter night when firewood brought in at bedtime still had snow on it the next morning.

I still live in an old home, but I've worked hard to insulate and weatherize it against cold Vermont winters. I also spent a lot of time choosing the right heating system to make certain it's a lot more comfortable than my boyhood home.

This article should help you to establish priorities and figure out what questions to ask your heating-system designer or contractor when choosing a heating system. I've tried to provide a broad overview, addressing basics of selection and fundamental decisions about heat-distribution options. Along with providing comfort, other goals that may come up when selecting a heating system include efficiency and size (sidebar, p. 53).

When it comes to heating-system design, there are two primary areas of decision-making: how to produce heat and how to move the heat where it's needed. The latter decision is usually more important and should be made first.

Forced air heats the fastest—Forced air is the most common type of heating system in North America. It accounts for 62% of all home heating, according to the U. S. Department of Energy. A gas-fired or oil-fired furnace (photo facing page), an electric furnace or a heat pump is used to generate hot air, which is then circulated throughout the house via ducts and a fan (air handler). Warm air typically is delivered through

Common heating terminology

Btu: **British thermal unit. One Btu is the amount of heat it takes to raise the temperature of 1 lb. of water 1°F at a specified temperature. The energy content of various fuels differs. Heating oil contains 139,000 Btu/gal., and propane has an energy content of about 91,300 Btu/gal. Kerosene has an energy content of about 135,000 Btu/gal. Natural gas is metered in units of hundred cubic feet (ccf) or therms (1 therm equals 100,000 Btus). A cubic foot of natural gas contains 1,027 Btus.**

AFUE: **Annual fuel-utilization efficiency. AFUE is a measurement of the efficiency of furnaces and boilers, averaged over a heating season.**

SEER: **Seasonal energy-efficiency rating. This is the ratio of the seasonal cooling performance of heat pumps and air conditioners in Btus divided by the energy consumption in watt-hours for an average U. S. climate.**

EER: **Energy-efficiency rating. This is similar to the SEER used for air-source heat pumps and air conditioners, except that cooling performance is not averaged over a cooling season.**

HSPF: **Heating-season performance factor. This number is a ratio of the estimated seasonal heating output in Btus divided by the seasonal power consumption in watts. It's used for comparing air-source heat-pump performance.**

COP: **Coefficient of performance, which is the ratio of the amount of energy delivered to the amount of energy consumed.**

floor registers. Air returns to the furnace through centralized return registers in the house (drawing facing page).

Advantages of forced-air heating systems include rapid delivery of heat throughout the house and the potential for integration with other climate-control systems, such as ventilation, humidification, dehumidification, filtration and air conditioning. On the downside are noise from the air handler and ducts, leakage through poorly sealed ducts, increased air leakage in the house resulting from pressure imbalances and the potential for drafts generated by air circulation. Also, it is difficult to set up zones that allow you to heat parts of the house separately.

It's important to plan carefully for ducts during the design of the house. Registers should be situated and the blower sized so that homeowners will not notice airflow. In poorly insulated houses, this may be difficult because the system often has to be designed to throw warm air across a room to offset heat loss at windows.

Baseboard radiators take up a bit more room in the house—In baseboard hydronic heating, a boiler heats water and a pump circulates the water through baseboard radiators around the house (drawing p. 50). Heat is delivered through both radiation and convection.

The most common baseboard radiators really don't provide much heat through radiation. Most of their heat is delivered by convection. Heat delivery can be regulated to some extent by using controllable flaps or louvers. Some European-style low-profile radiators, such as those produced by Runtal (800-526-2621), operate at lower temperatures and deliver mostly radiant heat (photo p. 50).

Hydronic heating offers quiet operation, minimal drafts, less heat loss from the distribution system compared with forced-air heating (if a

Drawings: Christopher Clapp

A system of ducts

Heat is generated in the furnace, which can be oil fired, gas fired, electric or heat pump. The fan, or air handler, creates pressure that forces the warmed air through the ductwork and out registers located throughout the house.

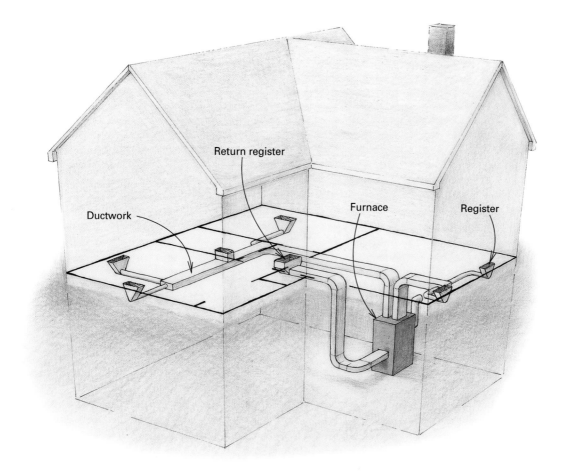

Return register

Ductwork

Furnace

Register

Forced air

Advantages
•*Rapid delivery of heat.*
•*Ducts also available for use with cooling, ventilation, humidification, air filtration.*

Disadvantages
•*Ducts can develop leaks, reducing furnace effectiveness.*
•*Fan and ducts make noise.*
•*Pressure imbalances can cause house to leak air.*
•*Difficult to zone heating within house.*

leak is present, you'll know it), no effect on air leakage from pressure imbalances in the house and ease of zoning different parts of the house. Negatives include high installation cost, interference with furniture placement if using conventional baseboard radiators and the inability to have the distribution system serve other climate-control functions.

The most subtle heat is from a radiant floor—Radiant-floor heating turns your floor into a low-temperature radiator. Typically, specialized plastic or rubber tubing is embedded in a concrete-floor slab, and water heated by a boiler is circulated through it (*FHB* #105, pp. 58-63) (drawing p. 51). Electric heating elements can also be used in radiant floors (*FHB* #75, pp. 68-72), but they are far less common and generally much more expensive to operate.

Radiant-floor systems are gaining in popularity because they provide a high level of comfort. Because of their comfortable average temperature and the fact that there is little stratification, homeowners are typically comfortable at air

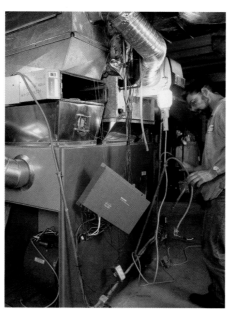

Forced-air systems offer varied climate control. Hot air delivered from this oil-fired furnace isn't the only use for a forced-air system. It's also used for air conditioning, ventilation, humidification and air filtration.

temperatures several degrees cooler than otherwise. This offers some potential for energy savings. On the downside, radiant-floor heating systems are expensive to put in, and they rely on skilled designers and installers who may not be available to work locally. Additionally, most radiant-floor systems respond quite slowly when the thermostat is raised because the whole mass of the concrete slab has to be warmed.

Direct heat originates at the point of use—Direct space heating is an alternative to the other distribution systems, which produce heat in one place then distribute that heat to where it's needed. Space heaters generate heat where it is used (drawing p. 52). Common examples are gas wall heaters, freestanding gas heaters, woodstoves, kerosene heaters, and wood-fired or gas-fired fireplaces.

Electric-baseboard heating is really a type of space heating in that the heat is generated at its point of use; but because baseboard-heating elements typically extend throughout the house, it is usually considered a whole-house heating

A system of pipes

In a house heated by hot-water radiators, the boiler heats the water to between 135°F and 180°F, well below the boiling point of 212°F required for steam radiators. Piping that is run through the house delivers the hot water to the radiators, which heat through radiation and convection.

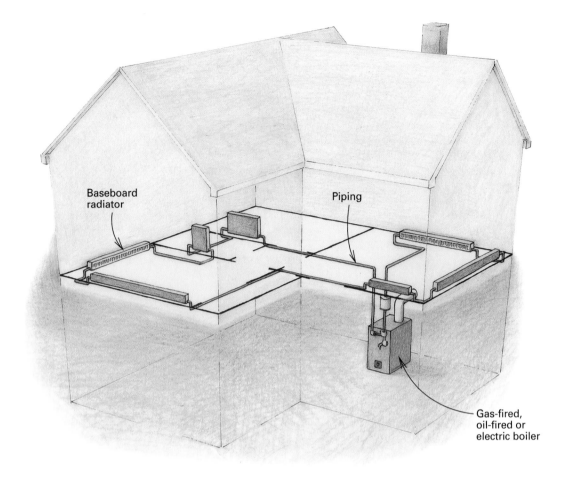

Baseboard radiator

Piping

Gas-fired, oil-fired or electric boiler

strategy. Electric baseboard heating also has a low initial cost and allows easy zoning. Also, there are no combustion byproducts, so there is no need for venting. Like hydronic-baseboard radiators, electric-baseboard heating elements work primarily by convection: Metal fins transfer electrically generated heat to air that circulates through the room.

A furnace or boiler is at the heart of most heating systems—The core of most heating systems is the centralized heating plant that generates the warm air or hot water that is circulated through the distribution system. With forced-air systems, a furnace or heat pump provides the heat. With hydronic heating and with most hydronic-floor systems, a boiler serves as the heat source.

Selecting the right furnace, boiler or heat pump from the wide range of products on the market can be difficult and confusing. The following guide to generic product types should help you to understand what's available. A discussion of space heaters, which are appropri-

Low-profile radiators need little space. Installed along the exterior walls, baseboard radiators use radiation and convection to heat a room. Air flows upward through the hot-water heated radiator, warms and rises to heat the room.

ate in some heating situations, is beyond the scope of this article.

Gas furnaces are among the most efficient—Induced-draft gas furnaces are what you'll find in most midefficiency models sold today. They use a fan to force flue gases either up a chimney or out through a side-venting pipe. Improved heat exchangers extract more heat out of combustion gases; as a result, the flue gases are less buoyant, and a fan is needed to exhaust them (hence the term "induced" draft). Because the flue gases are relatively cool, plastic pipe is sometimes used for side venting. AFUEs (see sidebar on common heating terminology, p. 48) are typically in the range of 78% to 85%.

Condensing-gas furnaces are the highest-efficiency gas furnaces and have such efficient heat exchangers that flue gases cool down enough for water vapor—one of the primary combustion products—to condense into liquid. When the water vapor condenses, it releases its latent heat, which boosts energy performance. AFUE efficiencies for condensing furnaces typi-

A system of tubing

A radiant-floor heating system is powered by a boiler, which raises the temperature of water to between 80°F and 140°F before sending it through the distribution system. The in-floor piping slowly raises the temperature of the surrounding material (concrete, tile, etc.) and helps to maintain a relatively even temperature.

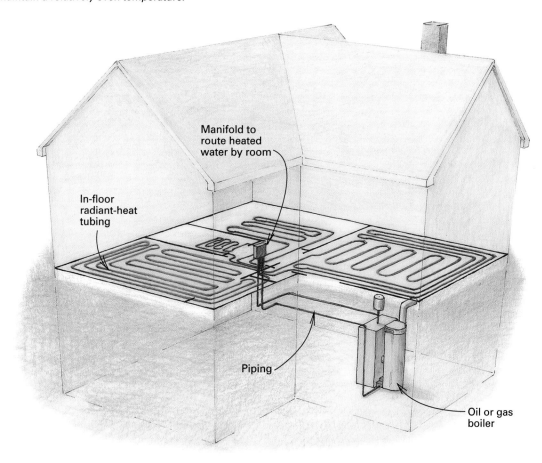

Manifold to route heated water by room

In-floor radiant-heat tubing

Piping

Oil or gas boiler

Radiant-floor heating

Advantages
•*High comfort level at lower temperatures, which saves money.*
•*Easy to zone house.*

Disadvantages
•*Does not respond quickly to thermostat changes.*
•*Relatively expensive to install.*
•*Requires designers and specialized installers, who may not be available in all areas.*

---

cally range from 90% to 97%. Flue gases are usually vented out through plastic piping. Condensate is piped to a floor drain.

Sealed-combustion gas furnaces force outside air into the combustion chamber and force exhaust gases outside; there is no need for or interaction with indoor air. Sealed-combustion equipment is considered the safest, with low risk of backdrafting. Sealed-combustion products are available in both condensing and noncondensing furnaces.

Like gas furnaces, gas boilers are available with venting configurations that achieve various efficiencies, but most are midefficiency induced-draft models. Practical efficiencies are somewhat lower with boilers than with furnaces because the temperature of the return water from the hydronic-heating loop is generally higher than the condensing point of the water vapor in the flue gases. That makes it difficult to squeeze out that extra efficiency represented by the latent heat of water vapor in the flue gases. Few gas boilers have efficiencies over 90%; most range from 80% to 85%.

Oil-fired furnaces have become more efficient. This oil boiler has an efficiency rating of 84% or better. Most oil furnaces and boilers on the market today have annual fuel-use efficiencies in the range of 78% to 82%.

Oil-fired furnaces and boilers are different animals—Oil combustion (photo left) is significantly different from gas combustion. Because the fuel is a liquid, it must be separated mechanically into tiny droplets and mixed with air for complete combustion. This is done with an injector-head burner that sprays air and atomized oil into the combustion chamber. Most oil burners today have flame-retention heads that increase turbulence in the combustion chamber to improve combustion. As with gas-fired equipment, oil furnaces and boilers built since 1992 have been required to have AFUE efficiencies of at least 78%. Most oil furnaces and boilers on the market today have AFUE efficiencies in the range of 78% to 82%.

Condensing-oil heating equipment is less common than condensing-gas equipment because of fundamental differences in the fuels. Oil combustion generates only half as much water vapor as natural-gas combustion, so there is less latent heat to be recovered from the condensation of the water vapor. Also, the condensate from oil combustion is much more corrosive than con-

Heat just where you need it

From direct-vented gas or kerosene heat to wood-burning stoves and fireplaces to plug-in electric baseboard radiators and heaters, space heating can be used to heat a whole house, a room or part of a room.

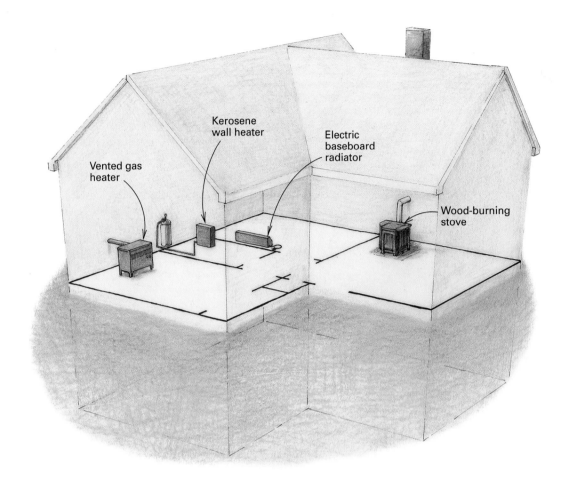

Vented gas heater

Kerosene wall heater

Electric baseboard radiator

Wood-burning stove

Direct-source heat

Advantages
- *Generates heat at point of use.*
- *Inexpensive to buy and install (except for those that require venting and flues).*
- *Easy room-by-room control of heating.*
- *In superinsulated house, may be cheaper than furnace or boiler.*

Disadvantages
- *Heats only small area.*
- *Doesn't perform other climate-control functions.*
- *Takes up space in rooms.*

densate from gas combustion. For these reasons, many heating experts recommend against condensing-oil furnaces and boilers, suggesting instead midefficiency models with AFUEs in the low 80s to the mid-80s. Few oil-fired furnaces and boilers are sealed combustion. In northern climates, mixing cold outside air with oil can reduce furnace performance or even cause start-up problems.

Electric furnaces are more expensive in colder climates—Electric furnaces are common in the southern United States, where heating loads are small and whole-house ducting is used for air conditioning. Low Btu (sidebar p. 48) output can be achieved without risk of spilling combustion products, and the heat can be readily distributed through air-conditioning ducts.

In all but the warmest climates, however, electric furnaces can be expensive to operate. Although the conversion of electricity into heat in the furnace is 100% efficient and a chimney or exhaust vent is not needed, electricity is usually

Fuel choices depend on where you live

Fuels are sold in cubic feet, gallons, cords and kilowatt-hours, and even when fuel units are the same, the energy contents differ. Some fuels burn more efficiently than others. Also, fuel prices differ from one region to the next.

When considering which fuel to use, it makes sense to compare them based on dollars per million Btus of delivered heat. First, you need to know how many Btus per unit of fuel there are and the efficiency at which you're burning that fuel. These values should be available from your heating-fuel supplier.

To compare the costs of various fuels, call the U. S. Department of Energy's Energy Efficiency and Renewable Energy Clearinghouse (800-363-3732) and ask for their brochure on the comparison of heating fuels.—*A. W.*

much more expensive than gas or oil, and losses in the duct system may reduce the overall heat-delivery efficiency dramatically. Also, unlike baseboard electric-resistance heating, room-by-room zoning is difficult.

If your heating season is short, an electric heat pump may be the best bet—Air-source heat pumps are amazing devices (drawing facing page). They extract heat from one place and deliver it to another. We're used to thinking of heat moving from warmer to cooler objects (for example, from inside your house in the winter to outside where it's colder). Heat pumps do just the opposite. Air-source heat pumps extract heat from the outside air—even if it's as cold as 30°F—and deliver it to your house, which may be at 70°F. The heat is delivered to the house as it is with a furnace—through ducts and registers.

Heat pumps accomplish this feat by circulating a special fluid called a refrigerant through a cycle in which it alternately evaporates and condenses. In the process, the refrigerant absorbs and releases heat. The process is controlled by a compressor, which changes the pressure of the refrigerant, and by heat exchangers, which allow the refrigerant to absorb or release heat from the surrounding air.

One of the best features of heat pumps is that they can be used for both heating and cooling simply by reversing the evaporation-compression cycle. In the cooling mode, a heat pump takes heat from inside a house and dumps it outside. Air-source heat-pump performance is measured by calculating its heating-system performance factor (HSPF; see sidebar p. 48).

If outdoor-air temperatures in winter commonly drop below 30°F, the seasonal-heating performance of an air-source heat pump will be compromised. Air-source heat pumps are most practical for moderate to warm climates.

Ground-source heat pumps benefit from the earth's constant temperature—Ground-source and water-source heat pumps are different from air-source heat pumps because they extract heat from the ground or from groundwater rather than from outside air (drawing right). Temperatures 8 ft. to 10 ft. underground are fairly uniform year-round. During the heating season, the temperature underground is considerably higher than that of the outdoor air; during the cooling season, underground temperatures are considerably cooler than the outside air.

In a ground-source heat pump, water or an antifreeze fluid is typically circulated through tubing that is buried underground; heat is transferred from the water to the heat pump's refrigerant through a heat exchanger. A water-source heat pump typically circulates groundwater through a heat exchanger to transfer heat to the refrigerant. The result is much better performance than can be obtained when outside air is used as the heat source. With ground-source heat pumps, a coefficient of performance (COP; see sidebar p. 48) of 3 to 4 and summertime energy-efficiency ratings (EERs) as high as 20 can be achieved.

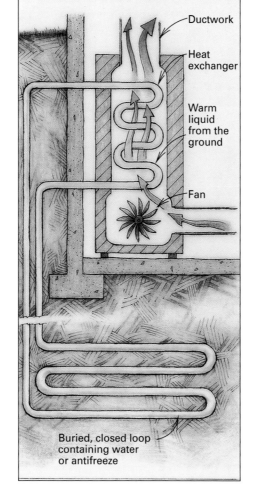

Heat pumps

In an air-source heat pump, outdoor air is drawn into the heat exchanger, which contains coils filled with a pressurized refrigerant gas. The gas, which in vapor form is at about 5°F, absorbs heat from the air. Next, the gas goes through a compressor, which raises the temperature before passing the gas through a second heat exchanger. The heat that is extracted is then forced through ductwork into the house. In a ground-source heat pump (below), water or antifreeze is circulated through underground tubing, where the heat-extraction process begins.

Ductwork

Heat exchanger

Warm liquid from the ground

Fan

Buried, closed loop containing water or antifreeze

The bottom line isn't always price or efficiency—When selecting heating equipment and components, specify durable products that will provide years of trouble-free operation. With major heating components, look for long warranties that are indicative of quality components and a long expected life. It is not uncommon for furnaces to have warranties of 20 years or more and boilers 30 years. Heat-pump warranties are shorter.

The equipment warranty doesn't tell you the whole story, however. Often more important is the availability of local service. "The warranty doesn't mean much if you don't have someone familiar with the equipment," notes engineer Mark Kelley of Building Science Engineering in Harvard, Massachusetts. Kelley recommends that his clients avoid heating equipment that isn't serviced locally, even if it is the most efficient or carries the longest warranty.

As you make decisions on heating equipment, don't lose sight of the fact that heating is part of an integrated system. The decisions that you make in designing a house and figuring out what construction details and insulation levels to incorporate will affect what type of heating system to use and how large it has to be. Also, don't forget that your overall goal with the heating system is to provide comfort, convenience, reliability and safety. The goal is not the heating system itself.

Before you begin thinking about what type of heating plant to purchase, decide on the heat-distribution system. Deciding how to get heat where it's needed is a more fundamental decision than how that heat will be generated. With a highly energy-efficient house, don't rule out the idea of using a space heater rather than a central heating system. ☐

Alex Wilson is coauthor of The Consumer Guide to Home Energy Savings *(American Council for an Energy Efficient Economy, 1996) and is editor and publisher of* Environmental Building News *in Brattleboro, Vermont.*

Choosing Ductwork

The kind of ducts you use and the number and types of turns they make dramatically affect HVAC-system performance

by John O'Connell and Bruce Harley

Think of air whooshing through ductwork as people rushing to work on a freeway. It doesn't take an accident to slow things to a crawl. Get thousands of cars traveling together at 60 mph, and watch what happens when the road curves.

Now suppose the highway has a 90° turn in it. Or picture an abrupt T-intersection where everyone must decide which way to go. Or think what would happen if one lane disappeared or if the road were filled with ruts and bumps.

Elbows, turns and intersections in ducts have the same effect on air. Every time there's an exit, the air slows down. Whenever there's a hill, air molecules pile up on each other and slow down, spreading the effect back down the line. If the ductwork is flexible, the ribs in the duct slow the air movement, just the way bumps in a road slow traffic.

This resistance is called static pressure. The type of ductwork that you use, the smoothness of its interior surface, and the number and severity of the turns and reductions the ducts make create this resistance. If ductwork is the appropriate type and size, and is as straight as possi-

It bends easily, but at a price. Flexible ductwork, such as this insulated, foil-covered duct, is good for getting up and over a joist or around a corner. However, the ridges inside the duct slow air so much that it shouldn't be used for straight runs.

ble, air moves with less resistance, and the air-delivery system performs better and with greater energy-efficiency.

Ducts to avoid, starting with flexible ducts—Although it's not possible to install all straight runs of round metal ductwork—which is the most effective method of moving air—it is possible to keep resistance to a minimum. One way to do that is to discourage contractors and homeowners from installing flexible duct.

Every straight foot of flex duct is equivalent to 2 ft. of rigid, which means the static-pressure penalty is twice as great with flexible duct. That's because the inside surface of flexible duct has the same effect on air as a rough road has on traffic. Add all the curves, uphill climbs and exits to a flex-duct system, and you can understand why it's difficult to get the performance you need from flex duct.

Originally designed to make short, vibration-dampening connections between branch ducts and register boots, flexible duct is now improperly used as an alternative to rigid ductwork. It is an attractive product. It bends easily (hence the name), goes around corners and travels up and

over joists. It's inexpensive, installs quickly and easily, and is available covered with insulation and a vapor barrier (photo facing page).

Even when used appropriately, however, flexible duct should be used as little as possible in air-delivery systems. And in reality, flexible duct often is not used appropriately. For instance, cramming it into a tight corner results in drastic airflow reductions. It is tempting to stretch flex duct too far. If it is, the strain may eventually cause the nylon straps that hold it to give way, resulting in a disconnection. If the disconnection is in an unheated attic or crawlspace, it might go unnoticed, which would mean a lot of warm, moist air pumped into a cold space. The outcome could be condensation, moisture and rot.

Also, if a lot of flexible duct is used in a system, the fan motor may need to be much bigger than would be necessary if rigid duct were used. In any case, the result would be either a system that can't deliver the air required or a fan that is oversize and, thus, costly to run.

Flex duct is useful in some circumstances—There are only a few situations where flex duct is worth the penalty. We use short pieces to get up

and out of joist bays, to connect branch lines to register boots or to get around some particularly tight corners.

Many equipment manufacturers recommend using a short piece of flexible duct or canvas boot connector where ductwork enters the heating or air-conditioning unit to break any vibration that might occur. Rigid-metal duct often transfers vibration through a system.

Some manufacturers also recommend using insulated flex duct in unconditioned spaces. They're concerned about condensation inside the metal ducts, which can lead to mold, mildew and bad air. They assume that the contractor won't remember to insulate the ducts in those spaces.

However, when use of flexible duct is considered necessary, it serves most efficiently if it's stretched to its fullest reasonable length. According to *ASHRAE Fundamentals* (one of a five-volume series of handbooks published by the American Society of Heating, Refrigerating and Air-Conditioning Engineers; 404-636-8400), flex duct at 30% compression increases the friction rate fourfold. If you absolutely need to use flex duct in a long run, make it one size larger than

Duct board is rigid fiberglass faced with foil. Preinsulated ducts made of duct board are generally leakier than rigid-metal ducts and puncture more easily. The seams are prone to failure unless assembled under clean shop conditions.

indicated and stretch it to full length—but don't pull it too tight.

Duct board and building cavities aren't the solution, either—Duct board is made of high-density fiberglass that has a foil-scrim kraft (FSK) facing, which is sort of an aluminum-foil version of rip-stop nylon (photo p. 55). Duct board is typically put together with FSK adhesive tape, which is prone to failure unless it's installed in clean factory or shop conditions.

Although it appears easy to create tight, preinsulated ducts using duct board, most of these systems are actually leaky once installed. Furthermore, the board's aluminum surface, which is the primary air barrier, punctures easily.

Stud spaces and joist bays are not duct runs, either, and should never be used as such—even for returns. First, they are nearly impossible to seal and to insulate properly. Second, introducing moist air to the framing of a house can't possibly be a good idea, as you can imagine. Blower-door tests indicate that many building cavities communicate with outdoor air much better than one might expect. In fact, many floor-cavity returns draw the majority of their air from the outside. Air traveling through a system tends to draw air with it through holes, leaks and seams.

The duct of choice is metal—Rigid-metal duct is the product of choice for air-delivery systems. Rigid ducting is the sturdiest and offers the least amount of static resistance (photo bottom right). Rigid metal also is resistant to accidental puncture.

By the time you've made the decision to use rigid-metal ductwork, you've already chosen the type and size of heating equipment you'll need and created a detailed line drawing that shows where the ductwork goes. From this, you can measure the length of duct runs and count the number and the types of pieces you'll use to put it all together, remembering to keep it as straight as possible.

Sizing the ductwork to the heating unit is a much more complicated process than we can get into in this article. Ductwork for heating and cooling systems must be designed based on room-by-room heat/cooling load calculations. Without this information, there is no way to know how much air needs to be delivered to each room. Part of the process of determining duct size entails figuring the total equivalent length (TEL) of each duct run in the system. Calculating the TEL of a duct run will give you a good understanding of what happens when you combine all those ducts and fittings (photo bottom left). For instance, a 90° square-to-rectangular elbow has an equivalent length of 35. This means that the resistance offered by this fitting is the same as the resistance of 35 ft. of straight duct. Once you get familiar with some of these numbers, it becomes easy to understand the penalties for a lot of corners, Ts and elbows, and easy to understand why particular rooms may not get the airflow that had been expected.

When you add up the equivalent lengths for all the corners, elbows, etc., you then can measure all the straight runs, which are called actual lengths. By adding the equivalent lengths of the fittings to the actual lengths of the straight runs, you get the TEL.

The size of the ductwork in a house is based on the longest TEL, according to the Air-Conditioning Contractors of America manual (ACCA; 202-483-9370). This means that some smaller duct runs may be oversize. This is fine in a ranch house. However, in a two-story house

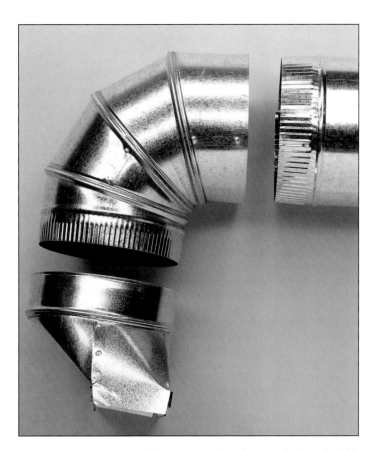

The best bet for ductwork. Round, rigid-metal ductwork is the best choice for delivering air throughout a house. However, these fittings—a 90° elbow and a round-to-rectangular transition—create resistance to the airflow and should be used judiciously.

Metal duct that doesn't need sealing. This ductwork made by Lindab (203-325-4666) is called SPIROsafe. Each fitting comes with a double rubber gasket at each end. The plastic sleeve connecting the two in the photo is for demonstration purposes.

or in a more complicated floor plan, you may have to size the ducts separately for each floor.

Needless to say, the longer the TEL, the more the ductwork resists air movement and the larger the blower has to be. And larger blowers cost more to buy and to operate.

Oversize ductwork is usually less of a problem than undersize ductwork. However, you don't want to oversize it greatly, especially in a low-velocity system such as a typical residential-ventilation system. It is important to go through the process of sizing your ducts correctly. Too big can be as bad as too small. In ducts that are too big, the fan blower does not have the ability to pressurize the entire system. Most of the air will exit the first available hole or grille, resulting in discomfort for rooms farther down the line.

Dos and don'ts of metal ducts—There are ways to use even rigid-metal ductwork incorrectly. Avoid excessively long runs and unnecessary turns and curves. In some cases long runs can be minimized by finding a better location for the equipment. Using rigid duct gets you thinking in straight lines and eliminates a tendency to go up, over and around.

When making corners, try to use longer smooth curves as opposed to 90° pieces (photo bottom right). And remember that Y-joints are less resistant than Ts. When installing a ventilation system, put the supply or the exhaust port high on the wall as opposed to the ceiling to eliminate a 90° turn in addition to a length of straight duct.

Finally, sealing and insulating ducts is extremely important if you want the system to perform as designed. Leaky ducts reduce equipment capacity, can cause backdrafting of combustion byproducts and generally cause air to flow in unintended directions. When a system is on, a leaky duct system can cause the air-leakage rate of the building to increase. Sealing and insulating ducts is more than we can get into in this article, however, and is worth an article of its own.

As usual, a little up-front work goes a long way—The most efficient way to deliver air is through a straight run of round metal ductwork, so layout planning is helpful. If possible, try to work with the builder before framing begins. If you get the builder to appreciate the issues that

you face either as a contractor or as a homeowner, the duct layout and system performance will be better.

For example, a good approach to improving the performance of duct systems in two-story houses is sending branch lines for the second-story rooms up through a center wall and then out through the floor to perimeter floor registers. This layout reduces equivalent lengths and duct size and eliminates some attic and exterior wall runs. This can be accomplished only if you can get the framer to put in a 2x6 interior wall somewhere and have the second-story joists lined up with it. Having the utility room located centrally may help to reduce duct runs. Working with the builder to plan for chases to carry ducts to distant parts of the house is also helpful in system setup. □

John O'Connell works for Energy Federation Inc. in Natick, Massachusetts. Bruce Harley works for Conservation Services Group in Boston, Massachusetts, as a technical adviser to the Energy Crafted Home and EPA Energy Star Home programs in New England. Photos by Scott Phillips, except where noted.

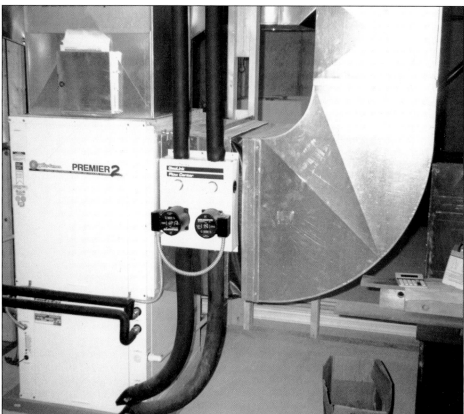

A curve is more efficient than a 90° turn. Air moves more efficiently through the large curve in this return duct than it would have if the installers had used a right-angle, or squared-off, fitting to bring air back to the heating unit.

Taking a Look at Windows

In a fog about choosing windows? You have to consider glazing systems, window styles and frame materials

by Jefferson Kolle

Part of my job as an editor is to read the hundreds of article proposals that arrive in *Fine Homebuilding*'s mailbox. For me, nothing sends up the red flag of rejection faster than seeing the word *fenestration* in a proposal letter. As in, "We tried to balance the subtleties of the *fenestration* with the implied massing of the built volumes."

Fenestration is a highfalutin word favored by architects and window manufacturers. It comes from the Latin word *fenestra*, which is the opening between the inner ear and middle ear.

Fenestration refers to the choice of windows in a wall or building. When choosing windows for a house, you'll have to decide not only the type of windows—casement, double hung, fixed, awning or slider—but also the kind of glass to use and the material used for the windows' frames. There's a lot of information, but it doesn't have to be confusing.

I've always thought that the most beautiful words—*fenestration* being an ugly one—are those that express ideas simply. So during the course of this article, I'll try to sort through all the window options without using the dreaded F-word.

Glass technology and window styles—Historically, the evolution of window styles has coincided with technological advances in glass-making. American colonists didn't use little panes of glass in their windows because they were trying to make their houses look like Ye Olde Gift Shoppe. Their windows had small panes because the technology of the day prohibited the manufacture of flat pieces of glass bigger than about 4 in. square.

As glass-making technology improved and the price of larger panes of glass got less expensive, windows with larger, fewer panes per window became common. In the early 18th century, it

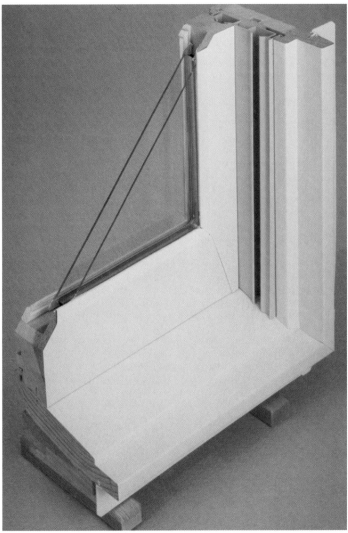

Insulating glass traps air between panes. Air is a poor conductor of heat and thus a good insulator. Insulating glass is made by sealing air between two or three panes of glass. The insulating glass shown here is separated by a stainless-steel spacer that is thermally isolated by polyisobutylene, which reduces conductive losses of heat and cold.

was not unusual for a sash to have 12 panes of glass, or lites, per sash. By the beginning of this century, one-lite windows were getting popular. Float glass, a new process developed in the late 1950s, makes it possible to produce pieces of nearly flawless glass in single-lite window sizes that are limited only by the size of the wall in which you want to install them.

Glass and energy-efficiency—In the 1970s, when everyone's energy-using consciousness was raised, people began to realize that most windows were nothing more than holes in the wall into which you pour heat. Storm windows had been in use for a long time, but storm windows are often inconvenient and ill-fitting, and they don't adapt readily to all window styles; for instance, you can't open a casement window more than an inch or two if there's a storm window attached to the frame's exterior.

Insulating glass—two panes of hermetically sealed glass separated by a spacer—was invented in the first quarter of this century, but it wasn't until the 1970s that the technology came into widespread use. Air is a relatively poor conductor of heat and cold, and having dead-air space between two pieces of glass works well to improve the efficiency of a window (photo left). The wider the airspace between panes—½ in. to 1 in. is common—the higher the insulating value of the insulating glass. Filling the space between pieces of insulating glass with gases that are less conductive than air—krypton or argon—further improves insulating ability.

Storm windows should not be confused with storm panels. Storm panels are tight-fitting pieces of glass attached semipermanently to a window's sash, where they don't impede the window's operation. Storm panels are not insulating glass; they are not hermetically sealed to the other glass in a sash. And a storm panel mounted over a piece of insulating glass should not be confused with triple-pane glass, which is true insulating glass with three pieces of glass sealed around two spacers.

Low-E glass can work in two ways—At the Massachusetts Institute of Technology (MIT), a group of scientists who later formed Southwall Technologies (1029 Corporation Way, Palo Alto,

Calif. 94303; 415-962-9111) developed the first heat-reflective low-emissivity glass coating. Glass with this coating is known as low-E glass, and it is an option offered by all window manufacturers. The physics of how low-E glass works is beyond the scope of this article, but what it does is relevant and important.

"Imagine a light bulb turned on," says John Meade of Southwall. "When you touch your hand to the light bulb, you are feeling heat through conduction: The heat is passing directly from the bulb to your skin. But if you hold your hand 1 in. away from the light bulb, you still feel the heat, radiant heat this time." Conducted heat can be slowed by insulation: fiberglass batts in a wall or dead-air space between pieces of glass. Radiant heat can be reflected away from the thing being warmed. This reflection is what low-E coatings do.

Low-E glass can work two ways, depending on which way it's facing in the sash. In a cooling environment—one in which you need air conditioning to keep a house temperate—low-E glass can reflect the sun's heat away from a house. In a heating environment, low-E can reflect heat (from your heating system) back into the house instead of having it pass through the windows. The direction in which low-E reflects heat is determined by which glass surface in a piece of insulating glass the coating is applied to.

Southwall developed another technology called Heat Mirror, which is a low-E film that can be suspended in the space between the panes in a piece of insulating glass. Tweaking the chemical makeup of the film's coating allows different types of Heat Mirror films to perform different functions. Some are better at reflecting exterior heat away from a house; others are better at keeping heat inside the house. All Heat Mirror films are superb at keeping ultraviolet (UV) light out of a house. UV light is what fades your furniture and carpets. Insulating glass with Heat Mirror film blocks 99.5% of the UV light that hits the exterior surface of the glass. Low-E insulating glass blocks only around 70% of the UV light.

Armed with insulating glass, triple-pane glass, low-E coatings, Heat Mirror films and exotic gases, window manufacturers and glass companies raced to see who could stuff the most efficiency-increasing components in a piece of insulating glass. A Canadian company, Willmar Windows (485 Watt St., Winnipeg, Man., Canada R2K 2R9; 800-665-8438) has come up with what it calls the R+12 glazing, which is two sheets of Heat Mirror film suspended in krypton gas between two pieces of glass, one of which has a low-E coating. Other companies have different systems for their most efficient glazing systems. As always, high technology and high R-values come with a high price (sidebar p. 62). Willmar R+12 glazing costs more than twice as much as regular insulating glass.

The metal spacer between pieces of insulating glass was a weak link in the efficiency of insulating glass. Metal spacers conduct the heat or cold that the insulating glass was supposed to block. Thus, the center of a piece of insulating glass might have had a pretty good R-value, but all around the perimeter where the pieces of glass

Casement windows swing like doors. Hinged on the side, casement windows swing past the plane of exterior walls. They can act as scoops to catch cooling breezes.

Six-over-six double-hung window pivots for cleaning. A handy feature of some double hungs is that the sash can be tilted into the room so that the windows can be cleaned from the interior.

Three window types ganged in a wall. Window manufacturers can gang different window types together in a wall. Shown here are trapezoidal fixed windows over casements over awning windows.

touched the spacer, the values fell way off. New designs have been developed that improve the cold-edge problem by isolating the metal spacer from the glass or by employing spacers made from nonconductive materials. Every window manufacturer will tout its spacers as the best ones on the market, and currently there is a this-year's-model immediacy in the hype that follows

a window manufacturer's improvement on edge-spacer technology.

Window styles are determined by the direction the sash move—Regardless of the season, my son's favorite bedtime story is *The Night Before Christmas*. About the 20th time I read the tale to him, he stopped me at the line "And threw

Top photo: Pella Corp. Bottom photos: Andersen Windows.

Energy-Efficient Building **59**

Sliding-window sash move horizontally. When they are opened, sliding windows don't protrude past a wall plane, making them perfect for locations where they open onto decks or other outside living areas. Sliding windows usually lift out of their tracks for easy cleaning.

Aluminum windows have thin frames. Because of aluminum's inherent strength, the frames and sash members are typically thinner than windows made of other materials.

open the sash," and he asked me what a sash was. I told him that a sash is one of the two basic parts of a window, the other part being the frame. I went on to explain that the window in the picture book was a casement window, but that I had seen other editions of the story where the window was a double hung. I was about to explain the difference between the casement window in the book and the double hungs in his room when my wife reminded me my son was only 3 years old and told me to keep reading.

All windows are similar in that their two basic components are a sash and a frame. The sash is the part of the window that holds the glass, and the frame is the part that holds the sash. Window types are differentiated by the method and direction in which the sash moves (or doesn't move, as in a fixed window) in the frame.

Double-hung windows consist of two offset sash, mounted one over the other in the frame. The lower sash slides up, and the upper sash slides down. In a single-hung window the upper sash is fixed, and only the lower sash slides.

The sash in a casement window pivots at the side of the frame; casements operate like little out-swinging doors (top photo, p. 59). French casements are two sash hinged on opposite sides of a single frame. Typically, French casements have no center-frame member between the sash.

Awning windows are like casements turned on their side so that the hinges are on the top of the frame, and the sash swing toward the exterior from the top. A hopper window is sort of the opposite of an awning window; the sash in hopper windows swing in, and they are hinged at the bottom. Sliding windows are like double hungs turned on their side; the sash slide horizontally.

Double-hung windows are the most popular—I don't spend a lot of time cleaning windows or even thinking about how easy it would be to clean a window were I so inclined. But window manufacturers give it a lot of thought. At a recent National Association of Homebuilders show, I tried the pivoting action on eight or ten windows on display. The sash on all the new double-hung windows tilted in for cleaning easily. In most cases, all you had to do was release a barrel-bolt-like clip on either side of the sash, and the window tilted right in. Some manufacturers use a compression jamb that allows you to pull the sash toward you easily (bottom left photo, p. 59). Andersen Windows Inc.'s (800-313-4445) double-hung windows even have a mechanism that holds the sash at a convenient angle for cleaning after you tilt it into the room.

One way to differentiate a double-hung window is to refer to it by the number of panes of glass, or lites, in a sash. A 12-over-12 window has 12 lites in the upper sash and 12 in the lower sash. Other common double-hung window patterns are 6-over-6 and 4-over-4. The thin pieces of wood that separate the glass in a sash are called muntins. Anything other than a 1-over-1 sash is referred to as having divided lites.

Divided lites and energy-efficiency—The energy crisis of the 1970s threw a curveball at the divided-lite, double-hung window. There was no denying the importance of energy-efficiency—all anyone had to do was to wait in a gas-station line—and there was no denying the fact that insulating glass was going to save homeowners on their heating and cooling bills.

The first insulated-glass windows typically had one large piece of insulated glass per sash. This design meant no divided lites. Architectural purists were in a quandary: how to get an efficient window and maintain a look they wanted.

Manufacturers came up with the snap-in grille. Snap-in grilles are grids—similar in appearance to a ticktacktoe setup—made of thin pieces of wood or plastic that snap onto the inside of sash and rest against the inside panes of insulated-window glass. There was one problem with snap-ins: They looked as fake as a cheap wig.

Another solution was to make sash that had individual pieces of insulating glass mounted in the rectangles formed by the muntins. But in order to cover the spacers that held apart the two pieces of insulated glass, the muntins had to be uncharacteristically wide. Muntins more than 1½ in. wide were common. Not only was the look inauthentic, but the wide muntins severely reduced the glass area in the sash. Another problem resulted from the increase in metal spacers themselves, which are the most inefficient area of insulated glass. For example, in a six-lite sash, there would be 24 cold edges—four around each piece of glass—instead of a single cold edge around the four sides of the sash's perimeter.

Manufacturers are constantly fiddling with the problem of true divided-lite sash. Some companies do offer sash with individual pieces of insulating glass captured by wood muntins, and some companies offer true divided-lite sash with affixed storm panels. But the more common method of dealing with the problem is to simulate the look. There are several different simulation tactics involving grilles that snap onto the interior of the insulating glass, ones that snap onto the exterior and ones that are sandwiched between the pieces of glass. Most manufacturers offer the previous solutions in combinations: You can order some windows with interior, exterior and sandwiched grilles.

Aside from the obvious energy advantages, another benefit of snap-in grilles (some grilles are permanently attached to the glass) is that some of them are easily removed when it comes time to paint the sash or clean the glass.

Divided-lite sash and simulated divided-lite sash are available from some companies in all their different window styles, not just in their double-hung styles.

Casement windows are hinged on the side—Casement windows were the dominant window style in this country and abroad until the beginning of the 18th century, when the single-hung window came into fashion. Today, casement windows are second in popularity only to double hungs. The hinge side of a casement sash pivots toward the center of the frame when it is opened. When the window is open, you can reach both sides of the glass for washing.

Typically, casements are opened by turning a small crank on the bottom of the frame. Their

Sorting through window ratings

Shopping for an energy-efficient window used to be like trying to buy a used car in a foreign country; you couldn't be sure whether you were being sold a lemon, and you couldn't understand the language. Some window manufacturers spoke of their products' insulation values; some spoke of air infiltration; some spoke of frame conductivity; and some spoke of solar gain. The result was an apples-to-oranges comparison that left you scratching your head.

The NFRC rates windows according to U-values—The National Fenestration Rating Council was started in 1989 to establish a standard energy rating for windows, doors and skylights. NFRC rates windows according to their insulating abilities. Independent testing facilities use computer modeling and actual laboratory tests using windows installed in wall sections to assign a U-value to a window. U-values are the inverse of R-values, so a lower U-value means a higher insulating value.

The NFRC is a nonprofit organization. Its revenues come from selling NFRC stickers with assigned U-values to window manufacturers that put the stickers on their windows.

Critics of the NFRC ratings say there is more to a window's energy-efficiency than its U-value. Air infiltration is important when judging a window's performance. A leaky window with a high insulating value is like a car that gets high gas mileage but leaks fuel all over the driveway.

Canadian ratings also consider air infiltration and solar gain—In terms of window ratings, Canada seems to be ahead of the United States. The Standard of the Canadian Building Code has established an energy-rating (ER) system that incorporates not only insulation values but also air infiltration and solar gain. ER ratings assign a number to a window. Windows are rated on a simple numerical system. A window with a negative ER loses energy, and one with a positive ER contributes energy. A window

with an ER of 0 is neutral in its energy consumption because it contributes as much energy through solar gain as it loses during a heating season. The shortcoming of the Canadian ER system is that it is only applicable to a climate where heating is the predominant energy cost; in the deserts of the United States, where air-conditioning costs are high and heating costs are negligible, solar gain is something to be avoided.

A new rating system is on the way—But take heart. The NFRC is working on a new system that will fill the needs of most all residential consumers. Brian Crooks is an NFRC researcher who works for Cardinal IG (7115 W. Lake St., Minneapolis, Minn. 55426; 612-929-3134), a company that has produced more than 500 million sq. ft. of insulating glass, According to Crooks, the new NFRC system will assign two numbers to a window: a fenestration heating rating (FHR) and a fenestration cooling rating (FCR). The numbers are based on calculations of a window's U-value, its air infiltration and its solar gain. Crooks said that "the numbers represent a percentage of total heating or cooling savings for one window versus another." So if one window is assigned a 10% FHR and another window has a 15% FHR, consumers will be able to tell at a glance that by using the second window they can expect an energy savings of 5% over that of the first window.

As of this writing, the new rating system is awaiting approval by the NFRC. Crooks said the new stickers with both FHR and FCR will be on windows early in 1996.

Between now and the time that the new NFRC ratings appear, your best bet is going to be to use the current NFRC U-value stickers and then spend a fair amount of time reading through manufacturers' catalogs, trying to sort out their convoluted test results. It's probably a good indication that if a manufacturer is forthcoming with its results that its window did pretty well in the tests. If no information is available from the manufacturer, you might ask why.—*J. K.*

ease of operation makes casement windows perfect for locations where sliding sash up or across is inconvenient if not difficult: over a kitchen counter, high up on a wall, etc.

Casement windows lock by means of small levers on the nonhinge side of the frame that clamp the sash to the frame. This clamping action makes casement windows highly resistant to air infiltration, and the harder the wind blows, the tighter the sash is pushed against the frame.

Taller casement windows usually have two lever locks, one above the other, and this upper lever can prove to be an impediment to people in wheelchairs. Peachtree (Box 5700 Norcross, Ga. 30091-5700; 800-477-6544) and Andersen offer optional hardware that allows the lower lever to control both locks. Taller casements made by Pella (102 Main, Pella, Iowa 50219; 515-628-1000) have two locks controlled by a single lever mounted low on the frame. Casement windows

Comparing window costs

The information in this chart is based on a double-hung window 2 ft. 8 in. wide by 4 ft. 6 in. tall with 1-over-1 sash and a jamb width of 4⁹/₁₆ in. Prices exclude screen, auxiliary muntins or options of any kind. Optional accessories add considerable cost to any window. Not all manufacturers make a window of these exact dimensions, so for comparative purposes, the windows closest to the stated dimensions were used.

The list prices are an average of the prices quoted from several lumberyards around Newtown, Conn., and from manufacturers. The prices are meant to be used as a point of relative comparison between windows made of different materials and using different glazing configurations. Window prices vary depending on place of purchase, number of windows purchased and other factors.

U-values are the inverse of R-values, so the lower the number the better. U-values in the chart were averaged from manufacturers' catalogs and from National Fenestration Rating Council literature.—*J. K.*

	Single pane*	Insulated glass	U-value	Low-E insulated glass	U-value	Gas-filled low-E insulated glass	U-value	Heat Mirror	U-value
Wood	$150	$210	.51	$235	.39	$250	.35	$301	.36
Vinyl	N/A	$225	.49	$250	.36	$260	.34	N/A	
Aluminum	$140**	$190	.74	$200	.62	$220	.59	N/A	
Vinyl-clad	N/A	$250	.51	$270	.39	$280	.35	N/A	
Aluminum-clad	N/A	$260	.54	$285	.41	$295	.36	$380	.39
Fiberglass	N/A	$360	.50	$376	.40	$390	.36	N/A	

* U-values are not available for single-pane glazed windows.   ** Aluminum-window prices and U-values are for single-hung windows.

can be hinged on either side. And when they are opened, casement windows swing past the plane of an exterior wall. For ventilation they can act as scoops to direct air indoors. Therefore, when you're ordering casement windows, it's important to know the direction of prevailing summer winds in your area because you can order your windows hinged on whichever side takes the best advantage of the natural convection.

Specialty windows—Installing an awning window under a large fixed window can provide ventilation. And placed high on a wall, awning windows can let in air and light while affording privacy. Because they swing outward (top photo, facing page), they can deflect light rain so it doesn't get into a building. Awning windows placed low on a wall can deflect winds hitting the side of a house up into a room.

Windows with horizontally sliding sash are called different things by different manufacturers: gliders, sliders, slide/bys. Both sash slide in a horizontal sliding window, and they usually lift out of their tracks for easy cleaning. On all horizontal sliding windows, the right-hand sash (viewed from the interior) slide on the inside track, and the left-hand sash slide on an outside track (top photo, p. 60). The sash slide past each other, but the window can be locked only with the inside sash to the right. Because they don't swing past the plane of a wall like casement windows, sliders are great for locations where you want a window facing a deck or outside space.

Inoperable windows are also referred to as fixed windows, and their shape is, as one catalog says, "only limited by your imagination." Manufacturers have thousands of sizes of fixed windows, and a lot of companies will make a fixed window in any shape you want. Aside from the more common rectilinear fixed windows, most companies have standard sizes of half-round, elliptical and trapezoidal windows. Trapezoidal windows often are installed so that their sloped side is parallel with the slope of a roof (bottom right photo, p. 59).

Because their sash don't open, fixed windows generally resist water and air infiltration well. Window manufacturers spend a lot of time perfecting weatherstripping around operable sash because this location is where water and air tend to invade. A fixed-sash window often is less expensive than an operable window of equal size. For a wall location where it might be hard to reach a window to open it, and where views and light are more important than ventilation, fixed windows can be a money-saving alternative.

Window catalogs are rife with photographs of huge walls of windows, walls that seem to be more glass than drywall. Ganging windows together is a common practice (bottom right photo, p. 59), and window manufacturers sell windows ganged in standard configurations. But windows don't have the strength of a stud wall; there are also wind loads to think about. If you envision a wall of windows for your house, it might behoove you to consult an engineer before you face the likelihood of getting turned down by a building inspector.

Wood windows need maintenance—In the earliest windows, metal was used to hold the glazing in place. By the 18th century wood was the most popular window-frame material, and today, wood windows still command about 50% of the residential-window market.

Until recently, wood has been a plentiful and relatively inexpensive material. And because wood is a poor conductor of heat, wood windows score high on energy-efficiency. But even though all of the parts of a wood window are treated with a preservative prior to assembly, wood windows require maintenance. In order to keep the windows looking good, you're going to have to get out your scrapers, putty and paint every couple of years and have a go at the exterior of a wood window. Some manufacturers will paint the exterior of your wood windows in the factory, and some of their paint jobs come with a good guarantee. Both Weather Shield (1 Weather Shield Plaza, Medford, Wis. 54451; 715-748-2100)

and Kolbe & Kolbe Millwork Company Inc. (1323 S. 11th Ave., Wausau, Wis. 54401-5998; 715-842-5666) warrant that their factory-applied coating will last ten years.

Clad-wood windows have a lot to offer—A fellow could get pretty hot and bothered when it comes time to maintain that "warmth and beauty" of his wood windows every couple of years. The perfect answer might be wood windows that are covered on the exterior with either vinyl or aluminum (bottom left photo, facing page). Many manufacturers make clad windows, and they do have a lot of advantages; you get a relatively maintenance-free exterior with a wood interior. Weather Shield even offers windows with oak or cherry on the interior

Although all windows with wood interiors are referred to as clad windows, cutaway photographs in manufacturers' catalogs—and all catalogs show cutaway views—show that there are different ways to clad a window. Some manufacturers attach the cladding by gluing it onto a wood frame. Others have designed their cladding so that it snaps onto a wood frame. Other companies make a vinyl or aluminum frame to which a wood interior is attached. Andersen has a unique system for its vinyl-clad sash: After the wood pieces for the windows are milled, they are covered along their length with a vinyl extrusion. The vinyl-covered pieces then are cut and fit into finished sash. Andersen Windows says that the wood adds strength to the vinyl. Andersen vinyl-clad frames are made by covering an assembled wood frame with an injection-molded vinyl sheathing.

Aluminum-clad windows don't suffer from the energy disadvantages of all-aluminum windows. That's because the heat-conducting properties of the aluminum are broken by the wood interior (photo bottom right, facing page).

The look of real vinyl?—A lot of people think of vinyl windows in the same way they think of vinyl siding: Vinyl is a material appropriate for a

Awning windows are hinged at the top and swing outward. Awning windows often are grouped below larger fixed windows to provide ventilation in a room. Marvin's Integrity windows, shown here, are wood, clad with fiberglass.

trailer, not a house. But vinyl windows are typically a lot less expensive than wood windows. And they never need painting. All-vinyl windows should not be confused with vinyl-clad windows. In the past, some vinyl windows experienced problems. Vinyl expands and contracts at a different rate than glass, and on some windows, repeated thermal cycling caused the vinyl to distort and to pull away from the seals around the glazing. Faulty seals can affect a window's ability to withstand air and water. Dark-colored vinyls absorb more heat than white or light-colored vinyl, and it was the dark-colored windows that had the most problems. One solution to combat the distortion/expansion problem is to form dark-colored vinyl over a light-colored vinyl core.

Vinyl is a brittle material that's 80% salt. In order to make vinyl pliable, a plasticizer is added so that it can be molded. With time, as the plasticizers evaporate, the vinyl gets brittle. New vinyl formulas, called uPVCs (unplasticized polyvinyl chlorides) are supposed to be more stable and resistant to distortion and movement.

An important feature to look for in vinyl windows is welded corners, rather than simple mitered-and-screwed corners. Vinyl corners are welded by heating both sides of a miter until they melt. The melted edges are pressed together; they then cool into one piece.

Vinyl windows are chasing hard at the heels of wood windows in popularity. Vinyl has always been popular for replacement windows, but the market for solid-vinyl windows in new residential construction is growing rapidly. Vinyl windows accounted for almost 16% of the windows sold for new construction in 1994. That number has risen from 3% in 1989.

Aluminum windows have gotten a bad rap—At least that's the opinion of Ralph Blomberg, president of Blomberg Window Systems (1453 Blair Ave., Sacramento, Calif. 95822; 916-428-8060). Blomberg points out that in recent years, some people have steered from aluminum windows because they are not energy-efficient. And some state energy codes have prohibited the use of some solid-aluminum windows. Aluminum is a good heat conductor and, consequently, a poor insulator. But, Blomberg contends, aluminum windows have some advantages, and they remain popular in the temperate West Coast climates. Aluminum windows are typically less expensive than wood windows or vinyl windows, and they shouldn't require much maintenance.

Aluminum can be painted, and a lot of manufacturers offer their products with a factory-applied colored coating. Another advantage of aluminum is that because of its inherent strength, you get a much larger glass area per window size (bottom photo, p. 60) than you can receive with vinyl or wood. And let's face it, you put a bigger window in a wall not so you can see more of the frame and muntins, but so you can look through more of the glass.

Fiberglass is not just for boats and surfboards—Odds are you've never seen a rusty Corvette or a rotten Boston Whaler motorboat.

Clad windows should never need painting. The advantage of vinyl-clad windows is that they have a wood interior and a maintenance-free exterior.

Aluminum on the exterior, wood on the interior. Wood's nonconductivity acts as a thermal break for the aluminum, mitigating the energy disadvantages of all-aluminum windows.

That's because both are made of fiberglass, and the stuff is almost impervious to the ravages of weather. Marvin Windows (P. O. Box 100, Warroad, Minn. 56763; 800-346-5128) recently introduced its Integrity line of windows clad with what Marvin calls Ultrex, a fiberglass material. And Blomberg Window Systems has recently introduced a line of all-fiberglass windows.

Fiberglass is not a new material, but only recently was the technology developed to pultrude fiberglass in the thin-walled, complex profiles needed for today's complex windows. Pultrusion is the process in which glass fibers are pulled through a resin bath and then into heated dies that cure the fiberglass.

Fiberglass has advantages as a material for both window frames and sash. Fiberglass expands and contracts at almost the same rate as window glass. This condition is advantageous because different degrees of movement during temperature fluctuations can cause the seals between glass and sash to rupture, which in turn can let in air and water. Other benefits are corrosion-resistance and dimensional stability; a fiberglass window on the south side of a house won't warp and twist. Fiberglass doesn't need much maintenance, although it can be painted. Also, the tech-

nology for pultruding fiberglass currently is limited to straight pieces.

Making an intelligent choice of windows—After you read this article, it's unlikely that you'll know which windows to buy for your next project. And that wasn't the intent any more than a general article about automobiles could tell you what car to buy next.

Before you decide on windows, do what you'd do before buying a car: Go to several dealers; take catalogs home and study them; and go back for another look. Are insect screens optional? Do you want to pay the extra cost of snap-in grilles or spend extra money for more efficient glazing?

Robert Wood at Hurd Millwork Company (575 S. Whelen Ave., Medford, Wis. 54451; 715-748-2011) made a telling comment: "When someone tells me that they don't want to spend a lot of money on windows, I tell them that windows typically account for between 3% and 5% of the cost of a new house. That's not a lot to pay for something you're going to look at and look through every day of your life." □

Jefferson Kolle is a former associate editor for Fine Homebuilding.

Top photo: Marvin Windows and Doors. Bottom left photo: Andersen Windows.

Energy-Efficient Building   **63**

Building a 'Green' House

Local limestone, ample overhangs and plenty of shade give this Austin home an edge against the Texas sun

by Peter L. Pfeiffer

The kitchen unfolds onto the screened-in outdoors. Sustainable design doesn't mean drab or plain, as this kitchen and its adjoining screened-in porch prove. When the weather is nice, the open porch door helps bring the outdoors inside, making this kitchen as comfortable and inviting as a spring day. Photo taken at A on floor plan.

A century ago, staking out a homestead in the rough and rocky central-Texas hill country around Austin demanded a common-sense person with a keen eye for the realities of land. The resourceful settler studied the slope of a site so that the layout and foundation of the house conformed to natural drainage patterns. The settler was sure to orient the house to take best advantage of the trees, the cooling summer breeze and the warming winter sun. And when it was time to build, the homesteader looked close to home for building materials.

The result was a rich heritage of such ranching homesteads—strong, simple forms constructed with walls of indigenous limestone and covered with tin roofs. Their owners expanded or added on to them as the needs of their families grew, so these old houses display an organic and logical progression of growth from an original main house to secondary winglike additions and even to barns.

My partner, Alan Barley, and I tried to think along these same common-sense lines as we sketched design possibilities for a house on a steep site in Austin. We wanted to minimize materials and costs, to work with the site and not against it, and to build a house that benefited from shade and prevailing breezes. We wanted to do all that and to give the house a sense of being well-grounded: We didn't want a house that looked like a box that had been airlifted in. The house we and our clients were after was basically what our ancestors had been after, only now such houses are called "green," or sustainable.

Green means committing to conservation and common-sense design—Green building starts with a design specific to the site and to the climate. In summer the house must benefit from prevailing cooling breezes and shade and during the winter respond to the sun for heat gain.

Resource efficiency is also a key element of green building (sidebar p. 67). It's more efficient to choose building materials that are produced locally and that, therefore, consume less energy to transport. In central Texas locally quarried limestone is a good choice over brick. Brick requires a high amount of energy to fire, must be trucked in from significant distances and, therefore, has a high "embodied" energy content.

Know what flows and what grows on the site—Understanding the site was our first step in controlling water and moisture problems. The building site is downhill from the street, so diverting runoff from both the front yard and the driveway was imperative. We accomplished this

A house that's armored against the relentless southern heat

Tall limestone walls shield against the Texas sun, a recycled tin roof reflects heat from living areas, and small windows on the sunny sides of the house prevent unwanted solar gain. The floor plan below shows more and larger windows on the eastern and southeastern sides of the house as well as a screen porch on the eastern side and covered deck on the southeastern side, which catch prevailing breezes.

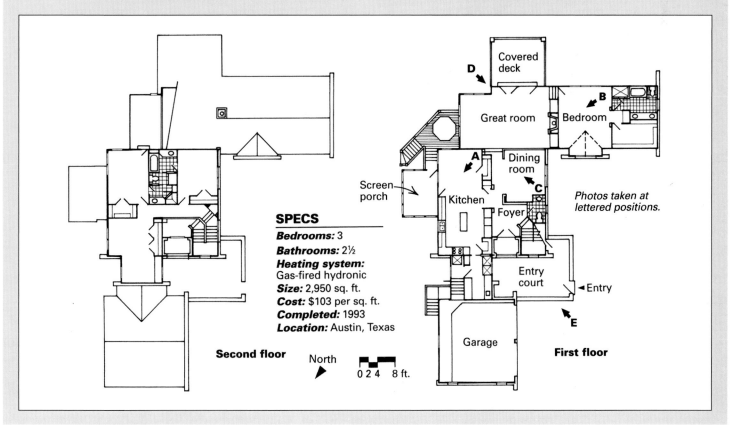

SPECS

Bedrooms: 3

Bathrooms: 2½

Heating system: Gas-fired hydronic

Size: 2,950 sq. ft.

Cost: $103 per sq. ft.

Completed: 1993

Location: Austin, Texas

Second floor

First floor

Covered deck

Great room

Bedroom

Dining room

Screen porch

Kitchen

Foyer

Entry court

Entry

Garage

Photos taken at lettered positions.

North

0 2 4 8 ft.

Drawing: Gary Williamson. Floor plan: Jeff Bellantuono.

Small, high windows cool the house in hot weather.
Acting as thermal siphons, small windows high on the north-
and northwest-facing walls draw out warm air and help to
cross ventilate the bedroom. Photo taken at B on floor plan.

Open interior spaces allow free flow of air. French doors lead to a covered porch, a wide-open living room opens onto a small dining area and piano alcove, and air circulates freely in the key living areas of this house. Photo taken at C on floor plan.

---

goal by setting the garage slab as high as we could without making too many steps down into the kitchen and entry court; and by placing a French drain at the garage foundation to divert groundwater to the sides of the house.

We also diverted driveway runoff from the house and garage by pouring a slight swale into the concrete drive, which directs water away, and by installing a French drain in the driveway in front of the garage doors. These steps had to be done subtly enough so that access to the house from the street wasn't awkward.

Of almost equal importance to understanding slope was the need to be aware of the type and location of major trees on the site. In our area, both evergreen and deciduous trees can help cool a house in warm weather. (Because of our mild Texas winters, collecting winter heat from the sun takes a backseat to the cooling benefits of shade and breezes.) While cedar trees can buffer the house from occasionally cold winter winds, their usefulness is greatest when they are thinned out to direct cooling summertime breezes toward outdoor living areas such as

decks and porches. Existing trees also can be used to enhance privacy. It's important to point out that the house on Dry Ledge Cove, where we were building, is in a closed-in and built-up subdivision of approximately ¼-acre to ½-acre lots. Regardless of lot size, privacy issues must be respected when designing a subdivision home that has large areas of glass. And privacy from the street was an important directive from the clients in their desire for a house that would be their retreat within the city.

Take advantage of the wind and protect from the sun—The hot Texas sun is a powerful force, and dealing with it takes more than oversize air-conditioning capacity. In the summer, unshaded western exposures take a beating from the sun (southeastern orientations are much better). Cooling summer breezes flow northwest from the Gulf of Mexico, so roof overhangs and covered porches on the southeastern sides of homes are effective. We calculated sun angles and the required size of overhangs to shade the windows in hot months using the Libbey-Owens-Ford sun-angle calculator (Libbey-Owens-Ford Co., Exhibit & Display Centre Inc., 1607 Prosperity Road, Toledo, Ohio 43612, attention William E. Baker; 419-470-6600; see *FHB* #80, p. 26).

The house is visually organized as several strong and distinct forms (drawing p. 65) with one "main" house, which includes the great room, and a series of additions: the garage at the front, the master-suite wing to the side and a covered porch to the rear. The additions' arrangement, the placement of windows and the size of overhangs along selected faces respond to the climate. Thanks to this organization, the major living areas—with their large windows and French doors shaded with overhangs and a covered porch—are protected from the hot afternoon sun and also benefit from balanced natural daylight, reduced glare and enhanced natural ventilation. Secondary service areas, such as the garage, laundry room, closets and circulation spaces, are placed on the opposite side of the house. These areas typically don't require as many windows, so they serve as buffers from the hot summer-afternoon sunshine or the occasional winter cold front from the northwest.

The relatively smaller windows on these north- and northwest-facing sides serve another important function. Light from windows high on these walls and in the stairwell provide for a balancing of daylight entering the house as well as a dynamic natural-lighting effect in the late afternoon. This design further connects the indoors with the outdoors without sacrificing privacy. These windows also help create thermal siphons within the house that aid cross ventilation. Breezes entering from lower southeast-facing windows rise up and out openings on the opposite (northwest) side. Therefore, the awning-type windows above the stair landing draw out air that comes into the main living areas through windows and doors on the southeastern wall.

Windows above the bed in the master suite (photo p. 66) provide for comfortable sleeping during spring and fall when natural ventilation provides an effective amount of cooling.

During warm or mild weather, two sets of French doors leading from the living room (photo p. 67) to the large and shaded covered porch extend the useful space of the main living area. The porch off the kitchen (photo p. 64) is an extremely comfortable, livable place, but more important, it doesn't require any air conditioning or heating.

Locally produced stone is a sensible choice for siding—Without knowing it, our central-Texas settlers used sustainable-building practices. Local limestone, cedar and tin roofs make a lot of sense. We applied a bit of modern-day physics to enhance the benefits of these local building materials.

We chose masonry to provide for a durable exterior veneer on the sides of the house that would be most subject to the brutal Texas sun. Limestone, which is abundant from quarries less than 15 miles away, remains unaffected by the drying effects of searing heat and was the logical choice for a masonry veneer.

We employed what around these parts is called the "German smear" technique in applying the mortar between and over the faces of the stones, which resurrected a visual effect produced more than a century ago for classic hill-country homesteads and commercial buildings (photo facing page). Windows and door headers project slightly from the face stone to aid in rainwater protection and to create a pleasing accent detail.

Interior applications provided another sustainable use for this stone. In the correct amounts, native limestone becomes an internal thermal mass that helps regulate indoor-temperature fluc-

Extra care ensures a long life for this cedar siding. Thorough application of a weatherizing stain on all sides and edges of this cedar siding keeps it looking good even after a couple of years of rain and sun. Two other details worth noting include the colored quarter-round molding that finishes the corner boards and the galvanized soffit material used as continuous, yet unobtrusive, crawlspace ventilation. Photo taken at D on floor plan.

tuations and keeps the house a little cooler in the summer and a little warmer in the winter. That benefit reduces the need for mechanical heating and cooling.

However, too much interior thermal mass can cause the house to take longer to cool in the summer or heat in the winter after being unoccupied for an extended period. To this point the client's lifestyle must be considered carefully.

The masonry fireplace between the living room and the master suite is placed on an interior wall (floor plan, p. 65) so that the radiant-heat effect of the stone, once warmed by fire, benefits both rooms. Also, less heat is lost to the outside than if the fireplace had been built on an exterior wall.

On the exterior exposures of the house, where the sun isn't so damaging, we applied cedar lap siding, which holds up well to decay, making it an economical and a durable choice.

Decker Ayers, our builder, had all the cedar siding and trim material unbundled to dry as soon as it was delivered to the job site. Then, with the siding and trim lying horizontally, he rolled Cabot's Weathering Stain and Preservative (Samuel Cabot Inc., 100 Hale St., Newburyport, Mass. 01950; 508-465-1900) on all sides and edges to maximize absorbency. This process was a bit labor intensive, but it made for superior protection and uniform appearance of the siding, even after it had been on the house for a couple of years (photo above).

Recycling and technology give the roof an edge over early counterparts—The tin roof is actually a series of 28-ga. channel-drain panels manufactured by Wheeling Corrugating Company (1134-1140 Market St., Wheeling, W. Va. 26003; 304-234-2400). We installed the panels in conjunction with a radiant barrier to produce an extremely long-lasting roof that effectively shields the attic from the summer sun.

This roof detail is an updated version of what local Austin settlers employed, enhanced by an understanding of modern-day low-emissivity coatings. The metal roofing is screwed over 1x4 battens placed 16 in. o. c. perpendicular to the roof rafters.

Thermo-Ply (Anthony Industries, Simplex Products Division, P. O. Box 10, Adrian, Mich.

49221-0011; 517-263-8881) nonstructural-grade reflective cardboard sheathing was placed over the rafters before the 1x4 battens were installed. With its shiny, low-E surface facing down, it becomes a radiant barrier that allows for convection air currents to exhaust heat out a continuous ridge vent before the heat penetrates the attic space. This design keeps the attic no more than 10°F warmer than the outside ambient-air temperature, so air conditioning and ducting operate in a cooler, less stressful environment. The R-30 blown-in cellulose ceiling insulation below, therefore, has less work to do to protect the living space from unwanted attic heat in the summer, substantially reducing utility bills and making for greater occupant comfort.

Sustainability is taken a step further. The galvanized metal roofing is made of recycled tin cans, automobiles and other postconsumer metal products; the reflective cardboard sheathing and the cellulose insulation are recycled-paper products. These steps alone won't save the world, but they prove that high-quality, durable components that contribute to a home's energy-efficiency can be made from stuff that a few years ago would have taken valuable landfill space.

The exterior-wall insulation, by the way, is certified "wet blown" R-13 cellulose mixed with boric acid for fire and insect resistance. Because of this material's ability to fill wall cavities effectively—especially those crowded with wires and pipes—air-infiltration levels for this house are low (below 0.65 air changes per hour).

High-quality windows, exterior housewrap, and extensive caulking and sealing throughout the framing, drywalling and trim stage of construction contributed significantly to the home's low air-infiltration levels.

Efficient mechanicals save money—Our house on Dry Ledge Cove employs some sophisticated yet simple HVAC equipment. Heating, hot water and cooling cost about $125 a month, which is about 35% to 40% of the expected average in this area.

The house was designed so that the second floor could be operated independently from the first floor and even shut down when not in use. The downstairs heating system is a "gas combo," or "hydronic," system that has been encouraged with financial rebates by the city of Austin's Green Builder Program. The water heater is in the crawlspace below the house and outside the indoor air-conditioned zone. When the house's thermostat calls for heat, the water heater circulates hot water through a coil in the air handler, which minimizes the introduction of combus-

Shading from the hot Texas sun. Plenty of shade trees and overhangs, thick masonry walls, small windows on the northern sides of the house and a recessed front door protect against the summer sun and provide privacy from the street. Photo taken at E on floor plan.

tion byproducts into the home and increases the water heater's efficiency by diminishing standby losses. It also has the added benefit of not drying the air, a problem standard to gas furnace, forced hot-air systems.

Both air-conditioning zones of the house are served by a single, residential-scale cooling tower manufactured locally by Allied Energy Corporation (1903 Westridge Drive, Austin, Texas 78704; 512-443-4466). This system uses evaporative cooling without incurring the high humidity that is usually associated with the standard "swamp cooler" design.

Typical outdoor compressor and coil units are replaced by coils in a cooling tower that constantly sprays them with recycled water. Evaporative cooling of hot refrigerant coils effectively doubles the efficiency of a standard, or air-to-air, system and remains highly efficient

even on the hottest days, when temperatures can exceed 100°F. This cooling process contrasts with standard units that are rated for 85°F days and whose energy-efficiency drops off precipitously during days warmer than that.

Water-conservation potential is built in—
This house is served by a reliable municipal-water system, so a complete rainwater-harvesting system wasn't economically feasible. However, without a lot of expensive, special construction, gutters and downspouts were arranged to make rain harvesting for landscape watering feasible in the future.

Landscape-watering needs, however, were kept simple and minimal through extensive application of the principles of xeriscaping, or landscaping with drought-tolerant plants. We carefully preserved the low-water-consumption plants

that were thriving on the site. Careful mapping of the existing trees and shrubs aided in designing a home around them; our environmentally conscious contractor kept the plants protected during construction.

The basics of a gray-water recovery system were included in the design and construction of this home—again, a simple, straightforward matter. Drain lines from all fixtures—other than commodes and kitchen sink—were run separately to the outside before being combined into one line to the city waste system. In this way, a future gray-water conditioning system and holding tank can be easily installed to allow for landscape watering with wastewater. □

Peter L. Pfeiffer is an architect with Barley & Pfeiffer, which specializes in green architecture, in Austin, Texas. Photos by Steve Culpepper.

Retrofitting a Threshold

A three-piece threshold provides extra weather protection, especially in exposed locations

by Gary M. Katz

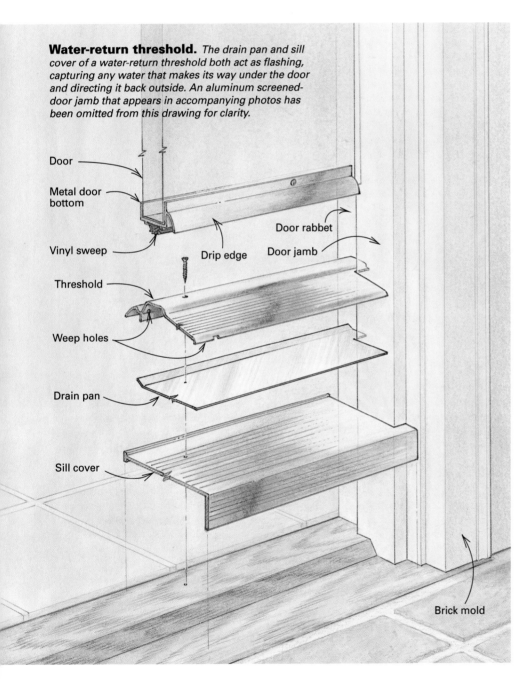

Water-return threshold. *The drain pan and sill cover of a water-return threshold both act as flashing, capturing any water that makes its way under the door and directing it back outside. An aluminum screened-door jamb that appears in accompanying photos has been omitted from this drawing for clarity.*

Door

Metal door bottom

Vinyl sweep

Drip edge

Door rabbet

Door jamb

Threshold

Weep holes

Drain pan

Sill cover

Brick mold

I used to install the ordinary type of metal thresholds available at hardware stores. Every time it rained, I'd worry. I'd worry about water sweeping in as the door swung open, water trickling in around the sides of the door or water entering through the screw holes. I'd worry about water warping a hardwood floor or staining a Persian rug.

Now I use ordinary thresholds only in protected openings. Experience has taught me that three-piece water-return thresholds are the safest bet. I often use thresholds made by Pemko Manufacturing (P. O. Box 3780, Ventura, Calif. 93006; 800-283-9988). A water-return threshold (drawing left) consists of a threshold, a drain pan and an interlocking sill cover. Although a water-return threshold is a little tough to install, the techniques I use make it simple enough, and the extra effort is worthwhile because it saves all of that worrying when it rains.

This example involves a door and frame already in place. While new, prehung-door units typically come with serviceable thresholds, the techniques discussed here could be used to add water-return thresholds to new doors.

Start with the sill cover—Sill covers are lifesavers. They are essentially the flashing for the threshold and cover the rough edge of a concrete slab or the exposed grain of a wood floor. They are also the perfect cure for elevation problems that can be created when, for example, a tile floor is laid right up to an original oak threshold and oak sill. This problem was the case in the door opening featured here.

I start by deciding how the sill cover should be notched around the jamb and exterior trim. Usually the sill cover butts against the door jamb or the brick mold (drawing left). On this job, a screened-door jamb had been added, and the sill cover had to remain behind that jamb so that the screened door would shut (photo 1).

Drawing: Dan Thornton

First, I cut the sill cover off square at the longest dimension needed, in this case from brick mold to brick mold. To cut the sill cover, I use a small circular saw equipped with a metal-cutting blade. (For more on cutting aluminum, see sidebar p. 73.) With the sill cover cut off square, I tip it into the opening and align it with the back of the screened-door jamb and with the rabbet for the main door. Using a pencil or a utility knife (scratch marks made by a knife are easy to see on most aluminum products), I scribe marks for the notch (photo 2). I repeat the process for the opposite side of the jamb. I'm using the jamb in place of a tape measure and square.

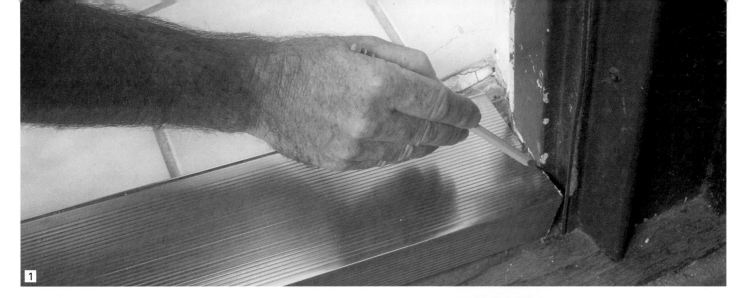

Notch the sill cover around the door frames. First, the broad, flat sill cover is held in place to mark the location and depth of the notch (1 and 2) that will allow the sill cover to fit around the jambs of the exterior door and the screened door.

Slope the sill cover to drain—After cutting the notches, I set the sill cover in place and prepare to trim the front, or vertical, edge of the cover. On some openings this step isn't necessary. But if there's a concrete porch or wooden step just beneath the sill of the door, then the sill cover has to be scribed in. The cover must fit tight to the original sill, and it must have some slope so that water will drain outside, not inside.

I tip the sill cover and check the slope with a torpedo level; between 1/8 in. and 3/16 in. of pitch across the width of the sill cover is usually enough (photo 3). Using anything handy, I shim the sill cover in place. Then, on the inside of the opening, I use my square to measure the distance between the sill cover and the floor beneath. I spread my scribes accordingly and scribe a line across the front of the sill cover (photo 4). Sometimes I attach a clean piece of masking tape to the sill cover to make the line easier to see. I put on my goggles and earplugs and, holding the sill cover as far from my face as possible, I cut to the line with my circular saw.

Start with the longest dimension—I start fitting the threshold the same way I fit the sill cover, by measuring the widest dimension of the

The sill cover slopes to the outside. The sill cover must be canted so that water can drain to the outside (3). Consequently, the front edge must be lowered by scribing it to the existing threshold, which is left in place (4). Masking tape makes the scribe line easier to see.

5

Scribe the threshold to the door frame. Rather than taking measurements and then transferring them to the threshold, the author holds the threshold itself against the jamb and carefully marks the locations of notches with either a pencil (5) or a knife (6).

6

7

8

Door height is transferred from the jamb to the door itself. Once the sill cover and threshold have been notched and set in place, the height at which the bottom of the door will be cut off can be determined (7). The mark made on the jamb takes into account the thickness of the metal-door bottom (8). The author then scribes the bottom of the door (9) to the new sill cover. Masking tape on the bottom of the door makes the pencil line easier to see.

9

door opening, the rabbet for the main door. After making the first cut for overall length, I slide the threshold into the opening to mark the notch (photo 5).

Normally, the threshold aligns with the face of the door, but for this opening I wanted to pull the threshold inside the house ¼ in. so that it would cover the raw edge of the tile. I tip the threshold, hold it against the jamb and mark the notches (photo 6). I repeat these steps for the opposite end and cut the notches.

Once the threshold is cut, I temporarily set it in place on top of the sill cover. I mark the spot where the front edge of the threshold rests on the sill cover. This mark will determine the location of the drain pan, which is installed between the threshold and the sill cover. It's important to locate the drain pan carefully so that it catches water seeping through weep holes in the threshold but at the same time remains hidden from view.

I set the drain pan in position just behind the mark on the sill cover, then scribe a line for the notch I need to make around the jamb. The thin drain pan is easy to cut with tin snips.

With the drain pan cut and in place, I set the threshold on top of it and drill pilot holes for the screws that hold the assembly to the floor. If I'm working on a concrete slab, I run my masonry bit through the threshold, drain pan and sill cover, down into the concrete. That's the surest method I know of getting concrete anchors in the right spots.

Cutting off the door—With the threshold and the sill cover in position, I'm ready to determine how much to cut off the bottom of the door. In order to get a weather-tight seal, this type of threshold requires a separate U-shaped metal door bottom with a vinyl sweep and drip edge (drawing p. 70); in this case I used one made by Pemko Manufacturing. In some installations, there is enough room for the metal door bottom between the bottom of the door and the new threshold. In this case, though, the door is too close to the threshold and has to be trimmed slightly. First, I measure up from the top of the threshold ½ in. and make marks on both jamb legs (photo 7). Then I remove the threshold and drain pan, but I leave the sill cover. The cover provides a smooth surface for me to run my scribes along.

I spread my scribes from the sill cover up to the line I've made on the jamb (photo 8), then shut the door and scribe a line across the bottom of the door (photo 9). Again, masking tape makes it easier to see the line.

I use different methods for cutting off doors. On veneered doors I sometimes use a "shooting

stick" straightedge made of thin plywood (photo 10). The shooting stick allows me to cut doors quickly without worrying about tearout. If I don't have my shooting stick with me, I use a metal straightedge and a knife to score a line on the door before I cut it. Either way, it's my circular saw that does the hard work.

Seal the threshold with plenty of silicone— Before I install anything, I sweep out all of the dust and dirt, especially under the sill cover. I run a bead of silicone under the sill cover to help secure it, though the screws that pass through the threshold really do that job. I press the sill cover into the silicone and then run another bead of silicone on top of the sill cover and beneath the drain pan. I take care to keep the silicone away from the front edge of the drain pan so that it doesn't squeeze out. Then I press the drain pan down into the silicone and apply more silicone at the joint of the jamb and drain pan. I also squirt silicone into the screw holes. Finally, I set the threshold and screw it down snug.

It's better to notch the metal door bottom around the weatherstripping on the door jamb, so I install the weatherstripping first if there isn't any already. The door bottom has to be cut to fit the overall width of the door and then notched to fit around the weatherstripping on the door jambs (photos 11, 12).

Not too tight, not too loose—After cutting the door bottom, I slip it on again and swing the door shut to make sure everything fits. The door bottom shouldn't be too long and squeeze or rub against the weatherstripping, but it should come close. I press the door bottom down against the threshold, but not too hard. The vinyl sweep needs to contact the threshold, but shouldn't be forced against it. Otherwise, the sweep will compress over time, and the seal will be lost.

I drive one screw at each end of the door bottom to hold it in place. Then I check the action of the door to be sure that it's sealing but that it's not rubbing too hard. Then I drive in the rest of the screws.

All that's left then is to apply silicone to the threshold and sill cover at the joint of the jamb, and maybe a little caulking between the door bottom and the door to seal out moisture. Before I leave, I check the swing of the door one more time. When it's right, the door closes just like a refrigerator, and with that whoosh of air, I'm gone. □

Gary M. Katz is a carpenter/contractor and writer in Encino, California. Photos by the author.

A site-built tool speeds the cut. The author cuts doors with a "shooting stick," a plywood straightedge that has a fence against which he registers the table of his circular saw (10).

First the weatherstripping, then the door bottom. Weatherstripping is installed on the side jambs first. The drip cap on the door bottom then is scribed and notched to fit around the weatherstripping (11 and 12).

Take care when cutting aluminum

Cutting aluminum thresholds is not my idea of fun, so I like to do it as quickly and as safely as possible. Many people use hacksaws. Some professional weatherstrippers use portable table saws with aluminum-cutting blades.

I use a 4⅜-in. Makita model 4200N trim saw (Makita U. S. A., 14930 Northam St., La Mirada, Calif. 90638; 800-462-5482) that cuts at a fast 11,000 rpm and a fine-toothed, combination metal blade, also by Makita (model 792334-2). I've used carbide-tipped blades to cut aluminum, but they are expensive, especially when the teeth begin to break off. (Makita no longer makes the 4200N, although some distributors still have a few of these saws. It was replaced with the model 5005 trim saw, which has a larger and slower-turning 5½-in. blade.)

Wearing eye protection is important when cutting aluminum, not only to guard against bits of flying metal but also because teeth on the combination blade can chip. In my work box I carry plastic goggles wrapped in a sock to prevent scratches. Years ago a bungee-cord accident took most of the vision in my left eye, so I'm careful with my right one. I also wear earplugs.—*G. M. K.*

A traditional profile that performs a modern function. This crown-and-dentil eave provides nearly as much ventilation to the rafter bays as a continuous aluminum soffit vent.

Venting a Traditional Eave

On a shingle-style house, aluminum soffit vents just won't do

by Robert Wills

The eaves on a house are like a handlebar mustache: a visual clue about the character lurking behind the detail. Whether a simple cove or a classical cornice, the eave detailing of a home helps to define its architectural signature.

But incorporating soffit vents in a traditional eave can present a challenge, particularly with a historic restoration or when trying to match an existing design on an addition to an old structure. Even venting a traditional profile in new construction can seem like an exercise in futility.

Old houses didn't need vents—When 18th- and 19th-century houses, with their delightful eave details, were originally built, insulation, dewpoints, vapor barriers and ventilation weren't issues. Without indoor plumbing and modern heating systems, old houses didn't encounter much warm, moist air. Any moist air generated in the house could escape through uninsulated walls and gaps in sheathing boards.

But today's tightly built, highly insulated houses are filled with warm, moist air, driven upward by physics, leaking into attics and roof cavities through any holes it can find. Once in the attic, the warm air can condense on the cold underside of the roof sheathing, causing mold, mildew and rot. Or the warm air can heat the roof and contribute to ice damming.

Although vapor barriers can help limit the flow of moisture into the space below the roof, ventilation—as provided by eave (or soffit) vents, gable vents and ridge vents—is still required to help protect a house's sheathing and framing

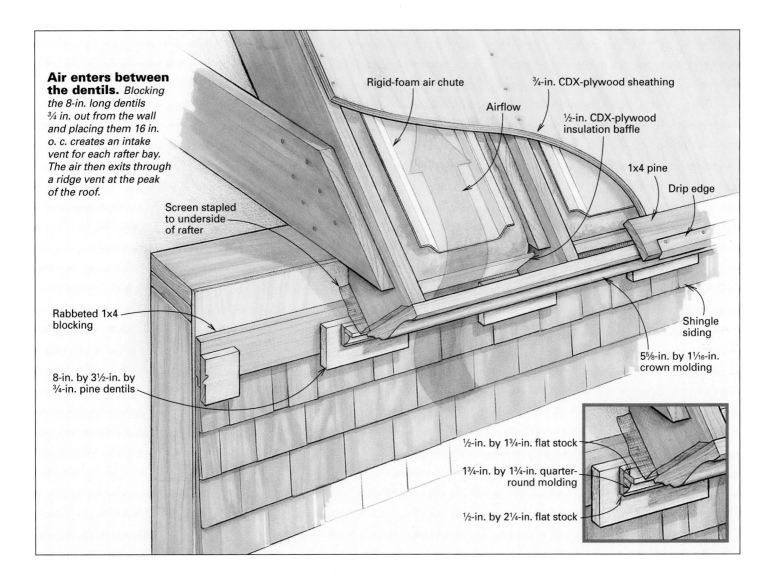

Air enters between the dentils. *Blocking the 8-in. long dentils ¾ in. out from the wall and placing them 16 in. o. c. creates an intake vent for each rafter bay. The air then exits through a ridge vent at the peak of the roof.*

Screen stapled to underside of rafter

Rabbeted 1x4 blocking

8-in. by 3½-in. by ¾-in. pine dentils

Rigid-foam air chute

Airflow

¾-in. CDX-plywood sheathing

½-in. CDX-plywood insulation baffle

1x4 pine

Drip edge

Shingle siding

5⅝-in. by 1¹⁄₁₆-in. crown molding

½-in. by 1¾-in. flat stock

1¾-in. by 1¾-in. quarter-round molding

½-in. by 2¼-in. flat stock

from excess moisture. Building codes require a net free-ventilating area equal to ⅟₃₀₀ of the total area of the space being ventilated if half of the venting is low and half is high. Because only 50% to 80% of the required ventilating area can be provided by ridge or gable vents, the remaining needs to occur in the eaves.

Dentils are the logical place to ventilate—
Camouflaging the modern technology of effective insulation and eave-to-ridge ventilation within the context of a historic solution to eave design can be tricky. Although the standard solution is to provide a gap in the soffit and then cover it with continuous aluminum soffit vents, this answer isn't always appropriate, particularly if there are no soffits to vent.

I was recently confronted with this sweaty problem while designing a new interpretation of a shingle-style house. Early practitioners of the shingle style, such as the firm of McKim, Mead and White, often incorporated classical motifs in their designs, such as ornate cornices. I, too, envisioned an eave with a combination of

classical trim profiles in my design. But how could a vent be placed unobtrusively in this assembly? I didn't know.

I started by making a section drawing through the wall and roof. After adding the line of the ceiling and required framing, I could quickly see that slotting or spacing any of the molding profiles in the eave assembly was unacceptable.

The final profile I arrived at for the eave (drawing above) was simply an assembly of stock moldings, including a 5⅝-in. by 1¹⁄₁₆-in. crown molding and a 1¾-in. by 1¾-in. quarter-round built up with ½-in. stock tacked to the back. Because the dentils were the only noncontinuous pieces of the assembly, they were the logical spot to insert ventilation. By holding the trim assembly ¾ in. farther out than I had planned, I could fur the dentils out, too, leaving the spaces between these "teeth" open to the rafter space.

Dentil size can vary, and it usually depends on the scale of the trim. Typical dentils for this size of house with this type of trim would be approximately 2 in. wide by 3 in. high by 1½ in. deep and spaced about 2 in. apart. I made long

and narrow dentils 8 in. long by 3½ in. high and spaced 8 in. apart. Although the proportions of the dentils might offend purists, this spacing enabled the vent openings to fall consistently on the open bays of the 16-in. o. c. rafters. It also visually simplifies the trim detail, appropriate with the shingled wall surface. This detail leaves each rafter bay with a ¾-in. by 8-in. ventilation opening. Although code would require 2.8-sq. in. per ft. net ventilating area according to the ⅟₃₀₀ rule, this detail provided 4.5 sq. in. per ft.

This eave detail requires a simple square-cut rafter tail, and for this profile there was no room for a 2x subfascia. Builder Greg Haeflin applied screening to the underside of the rafters before installing the trim, and he rabbeted 1x4 blocking at the top of the wall to hold the dentils out far enough to allow an air passage and to provide enough room to slide the top course of the shingles up and behind. Now, if I could only come up with a good ridge-venting detail. ☐

Robert Wills is an architect in Rhinecliff, New York. Photo by Andrew Wormer.

Understanding Energy-Efficient Windows

How today's high-tech windows work and what to look for when making your next purchase

by Paul Fisette

A respected builder I know told me how he learned the true value of energy-efficient windows. In the course of his business, he installed a builder's line of windows from a well-known manufacturer in every house he built. He felt good about his choice; he purchased the windows from a manufacturer with a reputation for quality, but they cost 10% less than the same manufacturer's standard line of low-E, argon-filled windows, saving him about $600 per house. He even put them in his own new home. The first winter he lived there, though, he noticed that the windows seemed cold. Only then did he compare the U-values with the same manufacturer's standard windows. He did some math and concluded that his windows were costing him about $150 a year. By his estimation, the low-E windows would have paid for themselves in four years and made his home more comfortable for their entire life span.

My friend based his conclusions on widely accepted averages, and although certainly not exact, they were probably not far off the mark. Experiences such as his are common, yet they are easily avoidable with a basic understanding of how energy-efficient windows work. When you choose new windows, appearance is often the first consideration. Initial cost is the next issue: Which window within the favored style costs the least? But liking a window's appearance is a fuzzy proposition, and cost really depends on durability and on the energy dollars pumped through the windows each year (chart p. 78). I am convinced that if we could see energy loss as we see color and shape, energy performance would top the list of window considerations.

Windows are thermal holes. An average home may lose 30% of its heat or air-conditioning energy through its windows. Energy-efficient windows save money each and every month. There are even some cases where new windows can be net energy gainers. The payback period for selecting energy-efficient units ranges from two years to ten years. In new construction, their higher initial cost can be offset because you'll probably need a smaller, less expensive heating and cooling system. And more-durable windows may cost less in the long haul because of lowered maintenance and replacement costs. Plus, you'll be more comfortable the whole while you live with them.

Keeping heat in (or out)—Windows lose and gain heat by conduction, convection, radiation and air leakage (drawing left, facing page). This heat transfer is expressed with U-values, or U-factors. U-values are the mathematical inverse of R-values. So an R-value of 2 equals a U-value of ½, or 0.5. Unlike R-values, lower U-value indicates higher insulating value.

Conduction is the movement of heat through a solid material. Touch a hot skillet, and you feel heat conducted from the stove through the pan. Heat flows through a window much the same way. With a less conductive material, you impede heat flow. Multiple-glazed windows trap low-conductance gas such as argon between panes of glass. Thermally resistant edge spacers and window frames reduce conduction, too.

Convection is another way heat moves through windows. In a cold climate, heated indoor air rubs against the interior surface of window glass. The air cools, becomes more dense and drops toward the floor. As the stream of air drops, warm air rushes in to take its place at the glass surface. The cycle, a convective loop, is self-perpetuating. You recognize this movement as a cold draft and turn up the heat. Unfortunately, each 1°F increase in thermostat setting increases energy use 2%. Multiple panes of glass separated by low-conductance gas fillings and warm edge spacers, combined with thermally resistant frames, raise inboard glass temperatures, slow convection and improve comfort.

Radiant transfer is the movement of heat as long-wave heat energy from a warmer body to a cooler body. Radiant transfer is the warm feeling on your face when you stand near a woodstove. Conversely, your face feels cool when it radiates its heat to a cold sheet of window glass. But radiant-heat loss is more than a perception. Clear glass absorbs heat and reradiates it outdoors. Radiant-heat loss through windows can be greatly reduced by placing low-E coatings on glass that reflect specific wavelengths of energy (drawing bottom right). In the same way, low-E coatings keep the summer heat out.

Air leakage siphons about half of an average home's heating and cooling energy to the outdoors. Air leakage through windows is responsible for much of this loss. Well-designed windows have durable weatherstripping and high-quality closing devices that effectively block air leakage. Hinged windows such as casements and awnings clamp more tightly against weatherstripping than do double-hung windows. But the difference is slight; well-made double hungs are acceptable. How well the individual pieces of the window unit are joined together also affects air leakage. Glass-to-frame, frame-to-frame and sash-to-frame connections must be tight. The technical specifications for windows list values for air leakage as cubic feet per minute per square foot of window. Look for windows with certified air-leakage rates of less than 0.30 cfm/ft^2. Lowest values are best.

Letting in the right amount of sun—In a cold climate we welcome the sun's heat and light most of the time. And once we capture the

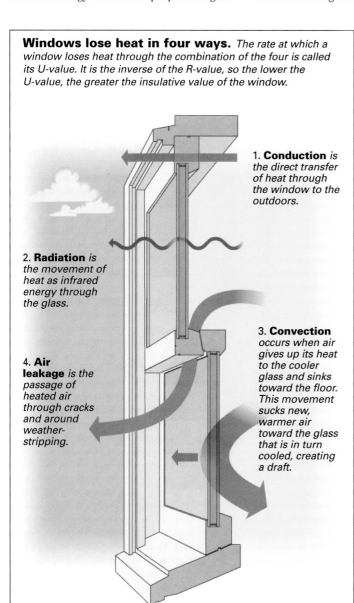

Windows lose heat in four ways. *The rate at which a window loses heat through the combination of the four is called its U-value. It is the inverse of the R-value, so the lower the U-value, the greater the insulative value of the window.*

1. **Conduction** *is the direct transfer of heat through the window to the outdoors.*

2. **Radiation** *is the movement of heat as infrared energy through the glass.*

3. **Convection** *occurs when air gives up its heat to the cooler glass and sinks toward the floor. This movement sucks new, warmer air toward the glass that is in turn cooled, creating a draft.*

4. **Air leakage** *is the passage of heated air through cracks and around weatherstripping.*

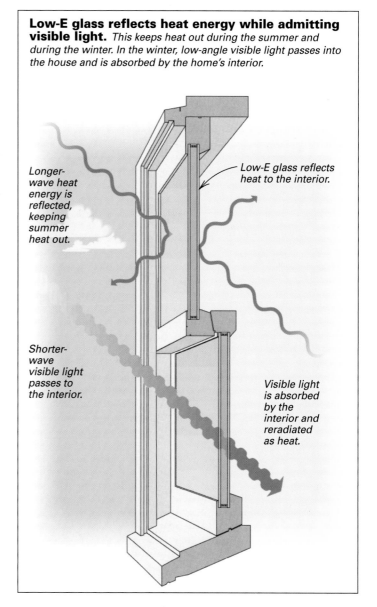

Low-E glass reflects heat energy while admitting visible light. *This keeps heat out during the summer and during the winter. In the winter, low-angle visible light passes into the house and is absorbed by the home's interior.*

Longer-wave heat energy is reflected, keeping summer heat out.

Low-E glass reflects heat to the interior.

Shorter-wave visible light passes to the interior.

Visible light is absorbed by the interior and reradiated as heat.

Window choice has a real impact on heating and cooling costs. *This chart is based on a computer model of heating costs for a 1,540-sq. ft. house with R-30 ceiling insulation and R-19 in the walls and floor. The window area is equal to 15% of the floor area.*

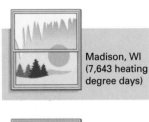

Madison, WI
(7,643 heating
degree days)

St. Louis, MO
(4,948 heating
degree days)

Phoenix, AZ
(1,444 heating
degree days)

Annual heating energy
(millions of Btus)

Annual heating cost
(based on an energy cost
of $6.40 per million Btu)

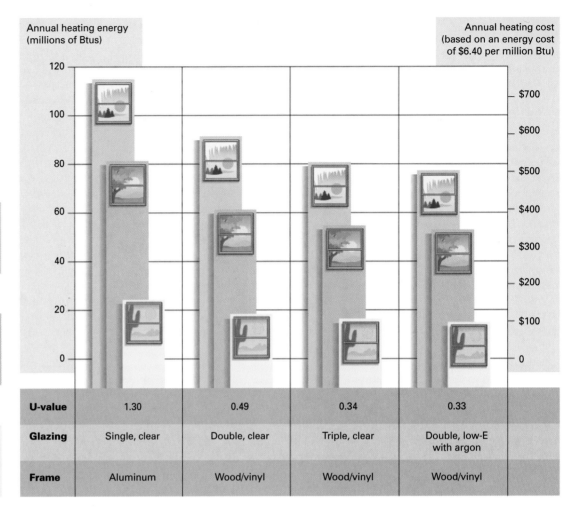

U-value	1.30	0.49	0.34	0.33
Glazing	Single, clear	Double, clear	Triple, clear	Double, low-E with argon
Frame	Aluminum	Wood/vinyl	Wood/vinyl	Wood/vinyl

heat, we don't want to give it up. In a warm climate, we don't want the heat, but we do want the light. Advances in window technology let us have it both ways.

Less than half of the sun's energy is visible (chart below). Longer wavelengths—beyond the red part of the visible spectrum—are infrared, which is felt as heat. Shorter wavelengths, beyond purple, are ultraviolet (UV). When the sun's energy strikes a window, visible light, heat and UV are either reflected, absorbed or transmitted into the building.

Only a fraction of the sun's energy is visible

There are windows that selectively block fabric-fading UV, visible light or infrared, which is felt as heat. Windows that block most UV and infrared while admitting visible light work well in cooling climates. For heating climates, choose windows that block UV while admitting heat and light.

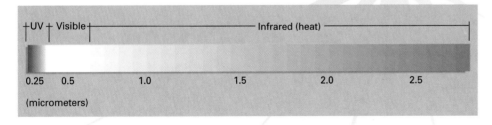

Enter low-E glass coatings, transparent metallic oxides that reflect up to 90% of long-wave heat energy, while passing shorter wave, visible light. In hot climates, they reflect the sun's long-wave heat energy while admitting visible light, thereby keeping the house cooler in the summer. And in cold climates, they reflect long-wave radiant heat back into the house, again while admitting visible light. This shorter wavelength visible light is absorbed by floors, walls and furniture. It reradiates from them as long-wave heat energy that the reflective, low-E coating keeps inside. Low-E

coatings work best in heating climates when applied to the internal, or interpane, surface of the interior pane. Conversely, in cooling climates, low-E coatings work best applied to the interpane surface of the exterior pane.

Low-E coatings improve the insulating value of a window roughly as much as adding an additional pane of glass does. And combining low-E coatings with low-conductance gas fillings, such as argon or krypton, boosts energy efficiency by nearly 100% over clear glass. Argon and krypton are safe, inert gases, and they will leak from the window over time. Studies suggest a 10% loss over the course of 20 years, but that will reduce the U-value of the unit by only a few percent. The added cost for low-E coatings and low-conductance gas fillings is only about 5% of the window's overall cost. It's a no-brainer.

Taking in the view—Windows with high visible transmittance (VT) are easy to see through and admit natural daylight. Besides giving you a nice view, high-VT windows can save energy because you need less artificial light. Some tints and coatings that block heat also reduce visible transmission, so be careful. Manufacturers list the VTs of windows as comparisons with the amount of visible light that would pass through

Chart information courtesy of W. W. Norton & Co. Inc., except where noted.

an open hole in the wall the same size as the window. VT is sometimes expressed as a "whole-window" value including the effect of the frame. What is important is the ability to see through the glass, not the frame, so be sure you get the VT of the glass, not of the entire unit.

The VT in residential windows extends from a shady 15% for some tinted glass up to 90% for clear glass. To most people, glass with VT values above 60% looks clear. Any value below 50% begins to look dark and/or reflective. Dariush Arasteh, staff scientist at Lawrence Berkeley Laboratory, warns, "People have very different perceptions of what is clear and what has a tint of color, especially when they look through glass at an angle." Look at a sample of glass outdoors and judge for yourself before you decide to order the window.

It's warm in the sun—Manufacturers have long used shading coefficient (SC) to describe how much solar heat their windows transmit. A totally opaque unit scores 0, and a single pane of clear glass scores 1 on this comparative scale. A clear double-pane window scores 0.84 because it allows 84% as much heat to pass as a single pane of glass.

Solar-heat-gain coefficient (SHGC) is the new, more accurate tool that is replacing SC to describe solar-heat gain. SHGC is the fraction of available solar heat that successfully passes through a window. It, too, uses a scale of 0, for none, to 1 for 100% of available light. The key difference is that SHGC is based on a percentage of available solar heat rather than on a percentage of what comes through a single pane of glass. It considers various sun angles and the shading effect of the window frame.

Glass coatings are formulated to select specific wavelengths of energy. It is possible to have a glass coating that blocks long-wave heat energy (low SHGC) while allowing generous amounts of visible light (high VT) to enter a home. This formulation is ideal in warm climates. A low SHGC can reduce air-conditioning bills more than if you increased the insulative value of your window with an additional pane of glass. I recommend a SHGC under 0.40 for hot climates. In cold climates you want both high VT and high SHGC. I recommend an SHGC of 0.55 and above in the North. In swing climates such as Washington, D.C., choosing a SHGC between 0.40 and 0.55 is reasonable because there is a trade-off between cooling and heating loads. For people in swing climates, Arasteh suggests, "Think about your specific comfort needs when specifying SHGC. If you like wearing sweaters and hate being overheated in the summer, then a low SHGC may be the choice for you." Choose the blend of glass coatings that works best in your climate and exposure.

Energy-efficient glazing reduces winter condensation

When low glass temperatures cause inside air to reach its dew point, water condenses on the window. The chart indicates the points where indoor humidity and outdoor temperature combine to cause condensation on various types of glazing. This chart is based on center-of-the-glass temperatures, but the edges are always colder, and condensation usually begins there.

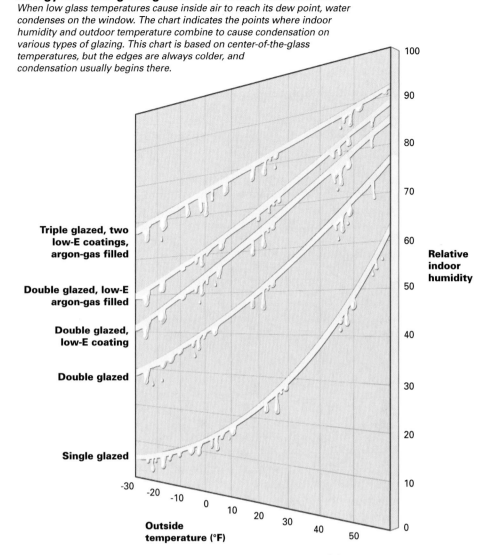

Preventing UV-damage—Windows that block UV-radiation reduce fabric fading. Expect to find windows off the shelf that block more than 75% of the UV-energy. Contrary to conventional wisdom, some visible light fades fabric, too. Some manufacturers use the Krochmann Damage Function to rate a window's ability to limit fabric-fading potential. It expresses the percentage of both UV and of that portion of the visible spectrum that passes through the window and causes fading. Lower numbers are better.

Window manufacturers sometimes boast R-8 (U-0.125) values. Be careful. This may be only the value at the center of the glass, which is always artificially higher than the whole-unit value. Look for whole-unit values of U-0.33 or better. Some manufacturers stretch low-E coated plastic film within the gas-filled airspace of double-glazed units to provide an effective third or fourth "pane." The weight of these windows is comparable to double glazing, and the true overall window performance is boosted to levels of U-0.17 or better for some. These units are pricey, but they can be more energy efficient than walls in cold climates. The R-value is lower than a typical wall, but if triple-glazed units are designed with a high SHGC and are placed in a sunny wall, they can be net energy gainers.

Keeping warm around the edges—If you've lived in a cold climate, you've seen condensation and even frost on windows. When warm indoor air cools below its dewpoint, liquid water condenses on the glass. Condensation typically develops around the edges of window glass. No surprise. The edge is where most multiple-pane glazing is held apart by highly conductive aluminum spacers.

The coldest part of a multiple-glazed window is around its edges (chart above). It's worse with true divided-lite windows; because each lite has edge spacers, the ratio of cold edge to warm center is much higher than with regular insulated windows. Moist conditions support mold growth, and hasten decay and paint failure. Condensation is the No. 1 reason for window-related

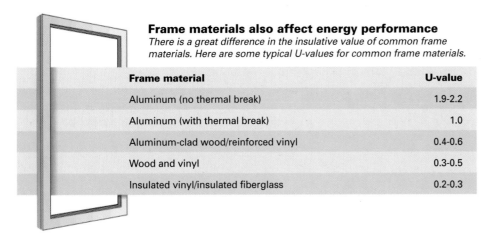

Frame materials also affect energy performance
There is a great difference in the insulative value of common frame materials. Here are some typical U-values for common frame materials.

Frame material	U-value
Aluminum (no thermal break)	1.9-2.2
Aluminum (with thermal break)	1.0
Aluminum-clad wood/reinforced vinyl	0.4-0.6
Wood and vinyl	0.3-0.5
Insulated vinyl/insulated fiberglass	0.2-0.3

callbacks. Warm edges reduce the chance of condensation forming.

The material the spacer is made from affects the rate that heat travels through a window's edge. Many window makers now offer warm edge spacers as standard fare. Aluminum spacers are not acceptable. The best windows use less conductive materials such as thin stainless steel, plastic, foam and rubber. Warm edge spacers can improve the U-value of a window by 10% and boost the edge temperature by around 5°F, thereby reducing condensation.

Good frames insulate—The most widely available window frames are wood (including vinyl-clad and aluminum-clad wood frames), with 46% of the market. Hollow vinyl frames hold 36% of the market, and aluminum runs a distant third, with a 17% market share. A trickle of alternative materials such as wood-resin composites, fiberglass, PVC foam and insulated vinyl makes

up another 1% of windows sold. A window's frame represents about 25% of its area. So it's important that the frame material be thermally nonconductive. For the most part, wood and vinyl are the best performers, and they work equally well (chart above). Aluminum frames are typically poor energy performers.

Connections where the frame joins together must be tightly sealed to keep out water and air. Weatherstripping needs to seal tightly after hundreds of window closings, rain wettings, sun dryings and winter freezings. Inexpensive, flimsy plastic, metal or brushlike materials don't last. Compressible gaskets like those used to seal car doors are best. Closures must clinch windows tightly shut. Look carefully at these components, and ask your architect or builder about a particular brand's track record. Pick longtime winners. Let others experiment with a new brand.

Wood is typically the most-expensive frame material. Maintenance is one of the biggest

drawbacks to using solid-wood windows. Wood rots, shrinks and swells. Paint fails. Solid wood requires frequent, fussy maintenance. On the other hand, well-maintained wood looks good, is stable and can be recolored easily. Clad versions are the easiest to maintain. On the down side, if you get sick of the cladding color, too bad. When you choose either a solid or clad version, be sure that the manufacturer has treated its wood frames with water-repellent preservative to improve durability, paint retention and dimensional stability.

Vinyl windows are built to move—Vinyl windows have been around for 35 years. Vinyl is energy efficient, durable, rotproof, insectproof and weather resistant. It's made with chemicals that inhibit UV-degradation. Vinyl is colored throughout and requires no painting. The knock on vinyl is that it fades, can't be painted, becomes brittle with age and is thermally unstable (especially dark colors). Temperature changes cause it to contract and expand more than wood, aluminum and even the glass it holds. Vinyl frames have the potential for causing increased air leakage over time because of this movement. But Richard Walker, technical director of the American Architectural Manufacturers Association (AAMA, 1827 Walden Office Square, Suite 104, Schaumberg, IL 60173; 847-303-5664), is quick to say, "Vinyl windows are built with this movement in mind, and failures have not been recorded to cause concern." If you choose vinyl frames, specify light colors and heat-welded corners. Heat-welded corners hold up best over time.

Rules of thumb for window selection
To get the best value from your windows, select units that match your climate. This chart suggests minimum values for the listed climates.

	Madison, WI (7,643 heating degree days)	St. Louis, MO (4,948 heating degree days)	Phoenix, AZ (1,444 heating degree days)
U-value	Less than 0.33	0.33	0.33
Visual transmission	50%	Greater than 50%	Greater than 60%
Solar-heat-gain coefficient	0.40-0.55	Greater than 0.55	Less than 0.40
UV-protection	Greater than 75%	75%	75%
Edge spacers	Warm edge spacers	Warm edge spacers	Warm edge spacers
Frame	Nonconductive frames	Nonconductive frames	Nonconductive frames
Air leakage	Less than 0.30 cfm/ft^2	Less than 0.30 cfm/ft^2	Less than 0.30 cfm/ft^2

1. U-values influence heat loss more in cold climates because the differences between indoor and outdoor temperatures are much greater than in hot climates.
2. Consider trade-offs involving comfort and performance in swing climates.

Bottom chart, this page: Information courtesy of Paul Fisette.

The pigments that are used in paint are almost identical to those used in vinyl, but vinyl's color goes all the way through. Walker says, "A little rubdown with Soft Scrub or one of the products on our (AAMA) list of recommended cleaners will bring vinyl back to its original brilliance." I tried the Soft Scrub test and was impressed with how much brighter aged vinyl became. Not the original color, to be sure, but the scrubbing resulted in a marked improvement.

Fiberglass-frame windows are showing up in a few product lines. Fiberglass is extremely strong, and because it is made of glass fibers, the frames and the glass expand at the same rate. Fiberglass must be painted and is more expensive than vinyl. Owens Corning, Andersen and Marvin are three major manufacturers that produce fiberglass windows. Owens Corning is the only manufacturer that makes fiberglass windows with insulated frames. But before you get too excited, the whole-window U-value for a low-E argon-filled casement window carries the same 0.32 rating for both an uninsulated vinyl and an insulated fiberglass unit.

Aluminum-frame windows are durable, requiring little maintenance. However, they are energy siphons and shouldn't be used where energy efficiency is a consideration.

The range of window options available today is staggering. But a working knowledge of the terms and these few guidelines should make choosing windows a little less intimidating (bottom chart, facing page).

You might also want to get ahold of this book: *Residential Windows: A Guide to New Technologies and Energy Performance* by John Carmody and others, W. W. Norton & Co. Inc. (1996). □

Paul Fisette is director of the Building Materials Technology and Management Program at the University of Massachusetts in Amherst, MA. http://www.umass.edu/bmatwt

Verifying energy performance

Until recently, purchasing a window was a little bit like buying a mattress. With mattresses you get cushion firm, chiro protector and Posturepedic Plus. Huh? With windows you're promised energy performance with U- and R-values. But what does the advertised R-value mean? And does it mean the same for all windows? Hardly. Some manufacturers determine R-value by measuring conductance at a single point in the center of the glass and do not count heat transferred through the frame or through metal spacers at the edges of the window. They do not account for the air that leaks around the sash. Nor do they measure radiant-heat loss from the entire window unit. Others honestly report whole-window values.

In 1989, the National Fenestration Rating Council (NFRC) was formed to level the playing field in the window industry. NFRC's mission is to develop a national energy-performance rating system for windows and doors (*FHB* #111, p. 38). All NFRC-rated windows are tested using a reliable, standard procedure that measures energy transfer through the entire window unit. U-values, solar-heat-gain coefficients, visible-light transmittance values and air-leakage rates are now (or will be soon) listed on certified windows. When consumers see an NFRC label (photo above) on the windows they are considering, they can be sure that they have a reliable tool they can use to compare windows.

As of June 1997, NFRC had over 150 participants with more than 30,000 windows, doors and skylights in the program. "This includes all of the major manufacturers like Jeld-Wen, Andersen, Marvin, Pella, Certainteed, Owens Corning, Peachtree, etc.," says Susan Douglas, the NFRC's administrative director.

How the NFRC system works—Window manufacturers that want to be certified hire a NFRC-accredited lab. The lab simulates the thermal performance of the windows with computers. Entire unit performance—including frame, spacer and glass—is measured. Windows with the highest and lowest simulated U-values in each product line (such as casements) are physically tested to verify the computer simulation. An independent NFRC-licensed inspection agency reviews the computer simulations and randomly pulls window units from the factory floor as test samples. Physical test values must fall within 10% of the computer predictions for a product line to be validated.

Presently, manufacturers that participate in the NFRC program must include certified U-values on the label. They also elect whether to include solar-heat-gain coefficient, visible-light transmittance and emissivity on the label. NFRC has created a technical procedure to measure air leakage and expected to have the details ironed out by January 1998. Douglas promises, "You can expect air-leakage values soon." Solar-heat gain is based solely on computer simulation, while visible transmittance, air leakage and emittance are physically measured values. The NFRC does not evaluate durability now, but the group does have a long-term performance subcommittee in place and expects to consider durability in the future.

Certified products have temporary and permanent markings. The big, temporary label is placed in a highly visible spot on the window. A small permanent NFRC serial code is etched on an inconspicuous part of the window, such as a spacer or metal strip. Permanent labels are useful. Potential buyers of older homes often ask, "What kind of windows are these?" A phone call to the NFRC will provide the brand and rating for the unit. Labels provide builders, designers, code officials and consumers with information needed to verify code compliance and a reliable level of performance.

Beyond labels—RESFEN, a computer program developed by Lawrence Berkeley Laboratory, enables you to minimize energy use, maximize comfort, control glare and maximize natural lighting in your home design. Builders, architects or consumers who want to fine-tune a design and choose specific windows for specific exposures on a house in a particular climate can order RESFEN from the NFRC for about $15. The NFRC Certified Product Directory and RESFEN are available from NFRC, 1300 Spring St., Suite 500, Silver Spring, MD 20910; (301) 589-6372.—*P. F.*

All NFRC-rated windows have this label. Manufacturers must list U-value, but other values are optional. Tests used to determine the values are consistent, so this label is useful for comparing different manufacturers' windows. Compare this window's rating with the chart at the bottom of the facing page. Photo courtesy of NFRC.

National Fenestration Rating Council Incorporated

AAA Window Company

Manufacturer stipulates that these ratings were determined in accordance with applicable NFRC procedures.

Energy Rating Factors	Ratings		Product Description
	Residential	Nonresidential	
U-Factor Determined in Accordance with NFRC 100	0.40	0.38	Model 1000 Casement Low-e = 0.2 0.5" gap Argon Filled
Solar Heat Gain Coefficient Determined in Accordance with NFRC 200	0.65	0.66	
Visible Light Transmittance Determined in Accordance with NFRC 300 & 301	0.71	0.71	

NFRC ratings are determined for a fixed set of environmental conditions and specific product sizes and may not be appropriate for directly determining seasonal energy performance. For additional information contact:

Building a Straw-Bale House

Made from agricultural waste, this low-tech, high R-value house satisfies California's tough building codes

by Janet Johnston and John Swearingen

Walls from fuzzy bricks. Building with straw bales is akin to assembling a wall with giant blocks. In the photo above, a bale is lowered onto a wooden frame that will eventually contain a window.

After the Wolf episode, the Three Little Pigs decided to rethink their construction methods and materials. They hit upon a bright idea. Why not build with straw bricks? They would have the insulating qualities of straw, the resilience of wood and the stacking qualities of brick.

The pigs were on to something. A house built with straw bricks—okay, bales—features a 2-ft. thick wall rated at around R-55 with seismic resilience, a two-hour fire rating and the kind of acoustic insulation that will muffle a leaf blower. Plus, a straw-bale house is just plain fun to build.

An ancient material comes of age—This article is about the first load-bearing straw-bale house to be built in California with a permit. When Carolyn North thought of building a vacation cottage (photo above) in Northern California's Andersen Valley, she envisioned a relaxed, informal house that would be equally comfortable during hot summers and gray, rainy winters. She and her husband, Herb, along with their co-owners, Daniel and Regina Schwenter, wanted an economical, energy-efficient house that would hold up in the event of an earthquake or a wildfire. They also wanted to participate in the construction to help keep costs down. Carolyn and her architect, Bob Theis, came to the same conclusion: Straw bales would be a perfect material for the cottage.

Even though straw has been used as a building material for thousands of years, it has only been since the turn of the century that mechanized baling machines have transformed a waste product—the stems left after harvest—into an annual crop of organic building blocks. We encountered plenty of skepticism to our project. Inspectors, plan checkers and worried neighbors all needed reassurance (sidebar p. 87).

Start with dry bales—A wet bale (above 19% moisture content) can succumb to fungal growth. So for openers, don't build with bales that have spent the winter uncovered. If they get superficially wet, bales can dry quickly. But once heavy moisture enters the center, rot can occur before the bale dries. We stacked our bales close to the building site and then covered them with a layer of black plastic, followed by a woven polyethylene tarp to protect them from the rain.

The potential for rot isn't the only concern about straw-bale buildings on people's minds. What about fire? Well, you could ask the same question about log houses. The truth is, the straw is so tightly packed that it's very tough to get it to burn, even harder when it's plastered.

Well then, won't the mice turn a straw-bale wall into their own private high rise? Straw doesn't have any food in it, so mice aren't interested in eating it. And its density and coverings discourage deep excavation. Truth is, for a mouse a fiberglass-filled stud wall is a much more attractive destination than a bale of straw.

Straw bales are just a wall system. Walls constitute about 10% to 15% of the total cost of the building, and a contractor-built straw-bale building will cost about the same as standard construction—except that with bales, you get a beautiful R-55 wall with plaster inside and out. Because bale construction is user-friendly, owner-builders can realize significant savings.

The walls can carry the load—Many bale buildings use bales as infill walls to enclose post-and-beam structures. But for this cottage, Bob and his associate, architect Dan Smith, decided to let the bales carry the load. There are advantages to eliminating the post-and-beam skeleton. For example, if you've got a wall that can carry the weight of the roof, why should you go to the trouble and the expense of building another structure to do the same thing? But in order to use the bale walls to support the roof, we had to reinforce the walls in a manner that

A load-bearing straw-bale wall
A broad footing with a capillary break of rigid insulation supports the wall. Threaded rods run from the footing through the bales, tying the wooden box beam atop the wall to the foundation.

2x4 load equalizer spreads roof loads.

Box beam

15-lb. felt

Stucco

Stucco lath

Tyvek

½-in. threaded rod

Straw bale

Wire ties, exterior to interior lath

Plaster over wire lath

Rigid insulation

2x4 plate

Radiant floor over rigid insulation

would resist the forces of an earthquake. More on this work later.

Bales come in various sizes. The three-string bales we used are about 16 in. by 24 in. by 48 in., although the bales in any batch will vary by a few inches. Our bales weighed 75 lb. to 90 lb., depending on density.

Our 15-in. wide footing flares out to form a 2-ft. wide base for the bales (drawing above). To keep the bales from wicking moisture from the ground, a capillary break is needed. We used rigid insulation for this purpose, flanked by a pair of pressure-treated 2x4 plates on the inside

and outside edges of the footing. The 2x4s eventually became nailers for the stucco and the plaster lath.

Bale raising is more than just building walls—An outfit called Out On Bale UnLtd. is the pioneer of the modern straw-bale movement (suggested reading, p. 86). Out On Bale is a resource center that has videos and books about building with bales. In addition, they hold seminars and workshops to get people pointed in the right direction. Out On Bale held a workshop during our bale raising, and friends of all ages and abilities showed up to help. After brief instructions to the helpers, the walls go up rapidly. The start of building creates a group euphoria that results in "bale frenzy," one of the dangers of bale construction. Adequate adult supervision must be provided at this time to avoid ending up with grossly out-of-plumb corners and missing window openings.

The first course of bales was "imbaled" on 16-in. long rebar stubs, two per bale, sticking out of the foundation. At about 6 ft. o. c., and around openings, we substituted ½-in. threaded rod for the rebar and ran the rod in 3-ft. sections linked by couplings all the way up to the beam at the top of the wall (drawing left).

Subsequent bales, laid in a running bond, were pinned to the courses below with a pair of 48-in. long rebar pins through each bale. In wheat and barley straw, the pins can normally be driven easily by a sledgehammer. Rice straw, which is denser, defeated our manly and womanly efforts to drive the rods. We resorted to inserting the rod with a ½-in. drill. The rebar then went in like a knife into butter.

When we had to cut a bale to fit, we first retied it to prevent the bale from unraveling. To pierce the bales, we used a "bale needle" fashioned from a ³⁄₁₆-in. steel rod. The needle has a flat, sharpened point on one end with a hole punched through the point. Baling twine was inserted through the hole, and then the needle was plunged through the bale at the intended cutline. We used polypropylene baling twine rather than baling wire because, though it degrades in sunlight, it won't rust and can be cinched tight using a trucker's hitch. Once these new twines were tight, the old twines were cut and the excess straw cut away.

Rice stalks being tough—almost 15% silica—cutting the bales became a major task. We tried hay saws, chainsaws, tree saws and knives of all descriptions. Amazingly, a serrated knife called a Ginsu knife (as advertised on late-night TV) worked best.

As walls grow higher, they become more wobbly. It is not recommended to make a wall over 12 ft. high or more than 30 ft. long without some

The windows are affixed to wooden frames. Architect Bob Theis contemplates the window bucks, 2x4 frames set in place as the walls are stacked. Bales were stacked against the bucks, and the bucks were then spiked to the bales with wooden dowels.

Wooden box beam tops walls. The box beam was constructed of 2x8s with ¾-in. plywood top and bottom (shown here without the top). All-thread rods extending from the foundation go between the doubled 2x crossbars.

buttressing. Plumbing and lining walls is easy, but corners are locked in as they are stacked and should be watched carefully during wall raising. Plumb corner boards are useful guides. Most errors can be easily corrected with a few kicks, a sledgehammer or a battering ram.

For window openings, we inserted 2x4 rough bucks in the walls as the bales went up (photo above left), pinning them with wooden dowels or spikes in the bales. The windows were nailed directly to these bucks. For straw-bale building, it is a good idea to make window and door openings oversize because the exact dimension of the bale opening after settlement is hard to predict. We made the window openings oversize by 6 in. to 8 in. in height, leaving a space for a block under the window. Header bales stretch over the tops of the windows.

Often a steel plate, angle iron or wood lintel is used to span openings, but in this cottage we used a continuous box beam atop the bale walls. The box beam acts as top plate, header, lintel and bond beam all in one (photo above right). It also was a platform for stacking the bales of our gable ends (top photo, p. 86).

Our box beam was 24 in. wide and was constructed of 2x8 sides and ¾-in. plywood tops and bottoms. This construction used a lot of wood, and it was also heavy and difficult to place. In our next house, we'll use full-length I-joists in place of the 2x8s.

Preloading the walls—An important use of the box beam is to provide a compression beam for preloading the bale walls. Traditionally, the bales are stomped into place as they are stacked, and then, after the roof is put on, the bales are allowed to settle for a few weeks before stuccoing. This step takes the initial fluff out of the bales, but accounts only for settlement and dead loading, not for any extraordinary loading such as earthquakes, wind or snow. By preloading the bales, we can ensure that no more settlement will occur during the life of the building. Compressing the bales has the added advantage of stiffening the wall and increasing shear strength. To discover how much the bales would compress under load, we erected a test wall (photo facing page) with water troughs hung from saddles resting over the top of the wall. We added water to reach 150% of ultimate design load (dead load plus live load) and found that bales could compress about ⅝ in. beyond initial settlement under extreme loads.

To link the box beam to the foundation, we ran ½-in. all-thread from the foundation through the center of the bales and left it protruding about a foot above the top bale. Then we lowered the plywood box beam, sans the top, over the bolts. Next we slipped some double blocks over the bolts and used hangers to secure them to the sides of the beam. We nailed the plywood top down and began tightening the bolts, using malleable washers to keep from crushing the plywood. We stopped after a couple of days of progressive cranking, having exceeded the required ⅝ in. of compression.

This technique had several drawbacks. For example, sometimes we stripped the threads. Impaling the bales on the bolts slowed the stacking process, and it was difficult to keep the bolts running straight through the center of the bales. In the future, we will run pairs of bolts along the outside surface of the bales. This process will make stacking easier, reduce the loads on the bolts, keep the bolts visible and accessible for replacement, and give us a way to take twist out of the box beam.

Engineering tests have been made on bale walls to determine compressive strength, in-plane lateral strength and out-of-plane lateral strength. To date, we do not have complete data on all aspects of bale construction, and the building department required us to demonstrate a means of handling overturning moment, the tendency of the wall to topple along its length. To handle this potential problem, we created shear panels, inserting cross bracing at points in the building (top photo, p. 86). This bracing consisted of steel straps bolted to the foundation and running to the box beam with 2x compression posts at the sides of the 2xs to complete the panel. We also placed trimmer studs next to door openings to avoid excessive point loading.

When we first thought of the box beam, we dreamed of a level platform for our rafters. However, because the object of the box beam is to compress the bales fully, the beam cannot be adjusted to level but must follow the contours of the bales. One of our first brushes with bale reality came at this time: the realization that this cottage would not be a dimensionally perfect building. The 1/16-in. increments on our tape measures looked pretty silly next to 3-in. variations in bales. On this job we wouldn't be demonstrating our skills in precision framing and impeccable finishes, so we ritually disemboweled our tape measures and began to concentrate more on building by sight, feel and common sense. For example, we framed the roof in conventional fashion but altered the depth of the rafters' bird's mouths by eye to fit the dips in the box beam.

Roughing in electrical and plumbing—

At the rough-in stage, plumbing and electrical lines can be run in grooves between the bales. Generally, these grooves are easily made either by clawing out with a hammer, a knife, a weed whacker or a chainsaw. But unlike stud construction, where it is easy to go through the top plate into another floor or area of the building, the bales and box beam can slow the tradesperson down. Careful thought at the planning stages can save a lot of head-scratching on the site.

Although some builders run Romex wiring in the walls, we instead elected to use metal flex (Bx cable). It's flexible, and its metal sheathing provides a level of psychological comfort. Outlet boxes are normally screwed to "vampire stakes," which are short, pointed 2x2s driven into the bales. The end grain of our rice-straw bales, though, was just too tough to take the stakes. We tried anchoring the boxes with wire threaded through the bales with the bale needle. But embedding a 16-in. length of 1/2-in. threaded rod into a predrilled hole and then bolting a metal box to the rod with a pair of nuts worked best.

Plumbing vents and drains are simply run through grooves cut in the bales, and again, the difficulty of penetrating the box beam should be anticipated. A 2-in. vent run up the side of the bale would hack through the 2x8 of the box beam, so the vent must be bent toward the center of the box beam.

We ran our water-supply pipes through the wall in plastic chases to guard against sweating

This is only a test. To determine the compression of a bale wall under severe loading, the crew built a test wall and loaded it with water-filled livestock troughs. With stucco in place, zero deflection occurred at a load of 600 lb. per linear ft. The troughs also provided a way for the crew to cool off at midday.

from condensation getting into the straw. We had no problem drilling through the bales with a sharp bit in our right-angle drill.

In some cases we furred out the bale walls, such as behind bathroom cabinets and where surfaces received tile. We were careful to provide wood blocking behind plaster surfaces to receive intersecting partitions, doors, cabinets and such. To hang electrical panels and equipment, we tied a plywood base through the bales, then stuccoed over it.

Controlling moisture in the bales—

The proper methods for waterproofing and vapor-proofing bales are a much-talked-about issue among straw-bale builders. Advocates of paperless building argue that paper can trap moisture attempting to escape to the outside and thus create a surface against which condensation can occur. If left to themselves, the bales would diffuse moisture out before the moisture reached a concentration high enough to support fungus; and even if rot began to grow (say, during a peak rain period), the rot would die as soon as the bales dried out. Any barrier to the free breathing of the bales thus becomes a threat to the health of the bales.

We think this approach is reasonable in mild or dry climates or where extreme conditions occur for short periods only. Even when paper isn't used to cover the entire building, a layer is

sometimes placed between the first and second courses of bales and hung over the bales to flash the first course. This paper protects the bottom course of bales from rain splash.

In climates where the outside walls might stay wet for long periods of time, roof overhangs will help if they don't inhibit air circulation. In very cold climates the possibility of vapor migrating from inside the house and condensing within the walls increases. This condition is especially true in walls next to kitchens and baths. In these conditions vapor barriers such as retardant paints may be used on the inside wall, as long as the outside walls can breathe. In very humid, warm climates, particularly with air conditioning, the construction should be reversed: vapor barriers on the outside and nothing on the inside, allowing the walls to dry to the interior.

Traditionally, plasters are applied directly to the straw, which provides an excellent mechanical bond. Code required that we use two layers of paper under stucco. We used a layer of housewrap followed by a layer of paper lath (bottom photo, p. 86). We also applied stucco wire, without building paper, on the interior to provide backing for exterior lath ties and to increase the shear strength of interior plaster.

We stapled the wire to the bottom plates and to the box beam and secured the wire on 12-in. centers horizontally and 8-in. centers vertically. But rather than slavish adherence to this grid, it's more important to make sure the wire is secured at the dips in the wall.

For anchors, we used 8-in. long "Robert" pins (too big to be called bobby pins), fashioned from 12-ga. steel wire. For maximum strength, we sewed the outside wire through the bales to the inside wire. We used two methods: bale piercing with hanger wires twisted to hook the wire on both sides; and continuous quilting with wire or twine threaded back and forth with a person on each side of the wall. Both methods worked to attach the wire firmly, and both methods required many hours of patient labor. We were thankful to be near a good radio station.

Builders often use expanded metal lath to create durable, rounded corners in plaster and stucco. To preserve the irregular shapes of the bales, we chose not to use expanded lath except where required for strength. For example, we used metal lath on inside corners to make the transition between partition walls and bale walls. We took great care in water-proofing around windows and doors, particularly on horizontal surfaces. Although we flashed openings according to standard methods used for stucco, the irregularity of the bales required meticulous care and patience to assure that no leaks would occur. On flat sills we used self-sealing bituminous sheeting in addition to regular building papers.

Before we plastered, we formed and poured windowsills in place with concrete made from white cement, sand and pea gravel. Once the concrete cured, we smoothed the sills with a sander. Although level, the sills conform to the irregular outlines of the window opening, and they make each window space unique.

Stucco emphasizes handmade appearance—One of the great things about bales is that they can be easily shaped with knives, saws or weed whackers. Before we began lathing, we carved niches in several locations. We also trimmed bales that protruded awkwardly and shaped our window openings.

To preserve the natural undulations of the bales and the rough, handmade appearance of the walls, we used standard stucco but applied only the scratch and brown coats. Some cracking appeared, but it isn't considered problematic. Cracking is typical with contemporary cement-based stuccos, which are very hard and brittle. To minimize the cracking, some builders have begun using traditional lime-based formulas, which are more flexible, have self-sealing qualities and allow the bales to breathe more than modern stuccos.

Inside, we used a stucco scratch coat topped with what is usually considered a base coat of plaster. We used both Structolite and Hardwall for the plaster coat (USG; 800-874-4968). Structolite dries white. Hardwall dries gray, and a little bit sandy. We found Hardwall a little bit easier to make smooth.

We omitted the third coat of plaster because we like the look of the base-coat material. It has a slightly rough appearance that preserves the soft, rounded shapes of the bales. But base-coat plasters are designed to be applied quickly, and thus have limited finishing qualities. They set quickly and cannot be troweled over more than once, and so require some skill and speed to apply over large surface areas. Plasterer Scott Greer and his crew looked panic in the face more than once when the plaster began to go off early, but they seemed to enjoy the whole ex-

Bracing at the corners. Steel straps running from the footing to the box beam were used to create shear panels. The panels include 2x4 compression members next to the straps; they work to keep walls standing in the event of an earthquake. The box beam over the bay-window opening serves as a 10-ft. header that supports the bales in the gable end.

The walls wear two wraps. The cottage was wrapped with Tyvek followed by stucco lath paper. To anchor the interior and exterior lath, the crew threaded wires back and forth through bales, passing the wire through loops in the stucco lath on both sides of the wall.

Suggested reading

Build It With Bales, A Step by Step Guide to Straw-Bale Construction **by S. O. MacDonald and Matts Myhrman.**

The Straw Bale House **by Steen, Steen and Bainbridge with Eisenberg. An extensive overview on straw-bale building with plans and drawings.**

These texts are available, along with other books and videos, from Out On Bale-By Mail, 1037 E. Linden St., Tucson, Ariz. 85719; (602) 624-1673.

The Center for Renewable Energy and Sustainable Technology maintains a web site with straw-bale information and a lively straw-bale list-serve at http://solstice.crest.org/efficiency/straw_insulation/index.html.

perience. They found plastering to the bales without paper much easier because the bales had less tendency to "rebound," accepting the plaster more easily than paper.

On the outside, we colored the stucco shortly after it had set by spraying on iron sulfate, a common garden amendment. The iron sulfate reacts with cement to make iron oxide—common rust—which permanently colors the surface. Great variation in colors from yellow to red naturally occurs following the countless natural variations in the stucco mix, such as temperature, degree of finish, proportions, setting time, curing conditions and who knows what. The result is deep, natural, attractive, interesting and rich colors that blend well with the cottage's rural setting.

We mixed the 5% iron-oxide salt with about equal parts water by weight and sprayed it (after straining) with a Hudson-type sprayer onto wetted walls. We also painted and rolled the solution in some places. The walls first turned a slimy green, but in a few hours, the color began to turn to a deep rust.

The revival of straw-bale building in the past decade has been a learn-by-doing movement. For us, innovations occurred regularly, leading to a lot of dialogue within the crew and with others around the country. Visitors came by daily and added input or took ideas with them. The technology is rapidly evolving, and at this point many techniques are borrowed from conventional building practices. As we gather more data about straw bales and have more experience building with them, methods and techniques specific to straw-bale construction are being developed (suggested reading left).

Simultaneously, the form and the aesthetic of straw bale is developing its own language, one of a soft, flexible material with qualities more like a basket than a wooden box. In these buildings, the walls themselves have a major presence. The thickness of the walls widens the threshold between the interior and the outdoors and changes the way light enters the rooms. The cottage we built has a deeply satisfying quality rarely seen in construction nowadays. It is a comfortable place.   □

Janet Johnston is a carpenter and architecture graduate. John Swearingen is a general contractor living in Junction City, California. Their nomadic design/build team works throughout the Western states. Photos by John Swearingen, except where noted.

Bale building takes patience and persistence

Building California's first load-bearing straw-bale house has been, for the most part, an adventure of invention, great teamwork and good fun. But as an owner/participant not normally involved in the building trades, I was not prepared for the amount of resistance, both bureaucratic and personal, that our project generated.

What surprised us was the initial climate of acceptance and interest, followed by months of balking and objecting to one point after another. We were promised a permit if we would provide extra cross bracing; then it was a test wall to be loaded with 7,200 lb. Then, if we wrapped the straw in an extra layer of moisture shielding, the permit would be ours. And when we agreed to comply with that, it was insisted that we come up with a flame-spread number—an impossibility, given that flames do not rush through material as densely packed as an airless bale of straw. At last, after consulting with every fire expert we could find and with other states' building departments, we were granted our permit.

But resistance also came from more unexpected sources. Some of the members of our own community, where the cottage would be built, succumbed to the fears of unknown technology. No amount of engineering data would convince them that we weren't adversely affecting their property values by building a house that would fall down, burn up or slump into a soggy pile of straw at the first big rain.

"Hold on until the house is finished," counseled architect Bob Theis during one of my darker moments. "The only convincing argument that you can come up with is the finished house, which will be gorgeous, cozy and totally at home with the landscape. You'll see."

And he was right. The finished house is captivatingly beautiful, tucked into the village meadow as if it had stood there for the past 200 years. And my community and, I hope, the building officials seem to be convinced, at last.

—Carolyn North, co-owner

Shedding Light on Skylights

New options in glazing, sizes and flashing systems can have you scratching your head when trying to choose the right skylight

by Roe A. Osborn

After 10 years my wife and I had almost gotten used to the dark, windowless crypt that served as our bathroom. There wasn't enough money in our original budget for a bathroom skylight, so we added it to our wish list, somewhere in between a new microwave and a garage. But after finding the right skylight at a sale, we moved the project up the list.

Using an old table lamp without a shade for light, I cut the hole in the ceiling and did the necessary framing alterations in the dim and dusty roof cavity. Then, on a steamy July morning, I stripped the roof shingles back and blasted through the sheathing, plunge-cutting with my sidewinder.

When I lifted out the rectangle of sheathing, a blast of air from inside the house blew sawdust everywhere. The change in the bathroom was dramatic. Even without the chase closed in, the bathroom went from dreary to cheery as sunlight flooded in. A refreshing, gentle breeze wafted through the room as the natural convection currents inside the house kicked in.

Skylights for different purposes. Skylights on the outside of this house punctuate the various roof planes. Inside, they serve a variety of functions, from kitchen light and ventilation, to ventilation and added views for a second-floor bedroom, to concentrated light and solar gain in the sunroom.

To vent or not to vent—Besides letting in light, our bathroom skylight was a way to get rid of excess moisture. One of the first things skylight buyers should consider is where the skylight will be going and if it will be needed for ventilation (photo facing page). If the skylight is going into a kitchen or a bathroom as ours was, a venting skylight is a wise choice. Even with exhaust fans and range hoods, a skylight can provide a quick alternate escape route for the excess moisture and warm air that these areas are likely to encounter (top photo).

Nearly every residential-skylight manufacturer offers the choice between skylights that open to provide ventilation and skylights that are fixed, or nonopening. Bart Mosser, vice president of Wasco (26 Pioneer Ave., Sanford, Maine 04073; 800-388-0293), told me that Wasco's fixed residential skylights typically outsell their operating skylights almost 3-to-1, probably because operating skylights almost always carry a bigger price tag. The venting version of their E-Class 22-in. by 46-in. skylight lists for $477 compared with $277 for the fixed.

Skylights in high-moisture areas such as bathrooms and kitchens are more likely to suffer condensation than skylights in other areas. Because venting skylights open to allow moist air to escape, they are best-suited for these high-moisture areas. If the condensation is heavy enough, it can run down the glass, over the skylight frame and down the skylight chase, damaging everything in its path.

Many manufacturers include condensation gutters on their skylights to catch and to collect condensation as it runs off the glass. If you are putting a skylight in an area likely to see a lot of moisture, make sure the skylight you choose has these gutters.

Still not convinced that you need a venting skylight? The folks at Velux-America Inc. (P. O. Box 5001, Greenwood, S. C. 29648-5001; 800-283-2831) have come up with a compromise. Their FSF skylight has a ventilation flap at the top of a fixed skylight (bottom photo). The flap opens into a channel to the outside that allows air circulation even in bad weather. The FSF skylight adds only about $30 to the cost of their FS fixed skylight of comparable size and glazing, a reasonable price for convenient ventilation. Velux, however, still recommends fully venting skylights for use in kitchens or bathrooms.

Motor-driven skylights open and close at the press of a button—If a skylight is being installed in a living space inside the geometry of the roof, such as a finished attic space, and you have easy access for opening the skylight, you may opt for one that closes with a latch and pushes open. Other skylights crank open and shut like awning windows. Push-open skylights open wider than their crank-out cousins and usually pivot for easy cleaning of either side.

If your skylight is going in a less accessible spot—say, in a cathedral ceiling high above the floor—then the crank-out variety is probably a better option. Many crank-out skylights also pivot for cleaning, but the opening mechanism has to be disconnected from the sash beforehand, not an easy feat when the skylight is out of reach. Instead of a handle, many crank-open skylights are equipped either with a socket or a small, fixed loop that lets you operate the skylight with a telescoping crank handle from the floor below.

Most skylight companies offer an optional motor that opens or closes the skylight at the push of a button either from a switch on the wall or a remote control. A lot of these motors look pretty ugly, like large boxes stuck on the skylight trim as an afterthought.

The slickest-looking mechanical skylight opener belongs to Roto's Sunrise II skylights (Roto Frank of America, P. O. Box 599, Chester, Conn. 06412; 800-243-0893). Roto houses the entire mechanical works in an aluminum extrusion that is wood-veneered to match the skylight trim. The aluminum extrusion conceals both the crank mechanism and the optional motor, and the extrusion can be removed easily for access and

A skylight expands a bathroom. The splayed chase for this skylight not only lets in more light but also makes the bathroom feel roomier. Even though the bathroom is equipped with a powerful exhaust fan, an open skylight can provide a quick escape route for the warm, moist air from the shower and the spa.

Fixed skylight with ventilation. Velux's FSF skylight does not open, but a ventilation flap at the top lets warm air out and cool air in through a screened channel. Photo courtesy of Velux-America.

A rain sensor waits for rain. When even a single drop of rain hits this small circuit board, it instantly signals the skylight motor to close the skylight and keep out the rain. Photo courtesy of Velux-America.

Smart skylights close when it starts to rain—When people give me a hard time about being bald, one of my standard comebacks is that you have to be bald to appreciate fully a ride in a convertible. Another advantage to being follicularly challenged is that I'm usually the first to feel raindrops. Some of the contractors I used to work for depended on me for this ability, especially when they were trying to get that last course of shingles on before an approaching thunderstorm.

One thing prospective skylight owners worry about is not being home to close the skylight in case of rain. You could hire a bald guy to sit on the roof and wait for rain, but skylight manufacturers have come up with a better idea: rain sensors, which are small printed circuit boards that look as if they shouldn't be exposed to bad weather (photo left).

The rain-sensor circuit board basically consists of two conductors in a grid pattern. When a drop of rain lands on the grid, it completes a circuit that signals a motor to close the skylight.

Virtually every skylight company that offers a motorized opener also offers a rain-sensor option for their operating skylights. The better openers have battery backups for their motors. You may consider battery backup to be overkill, but if the electricity is knocked out and it starts to rain (as often happens in thunderstorms), you'll appreciate battery backup.

Some skylights with motorized openers can also be opened manually in case of emergency. This feature can be important if the motor fails or if you need to open the skylight before the power comes back on.

The controls for automated-skylight functions vary among manufacturers. Some control just the opening and closing functions, and others can be programmed to control motorized-skylight accessories such as the interior shades that most companies offer. Velux says it has a system that will interface with any homewide computer system. The system can control multiple functions on multiple skylights from a single remote box.

Glazing options can reduce UV-damage and heat transferral—Jefferson Kolle, a former renovation contractor, told me that he once dropped a worm-drive circular saw onto a skylight of tempered glass and was amazed when the saw just bounced off. For safety reasons, skylights are made of tempered glass or laminated glass. Tempered glass is extremely strong, but when it breaks, it shatters into a million glass pebbles.

Laminated glass has a thin plastic sheet attached to the glass. It's generally not as strong as tempered glass, but when it breaks, the plastic keeps the glass in a sheet. For this reason, codes in some areas specify laminated glass for the interior-facing skylight glass in certain applications, such as over a bathtub or spa. Skylights with laminated glass usually have tempered glass on the exterior and are designated "tempered over laminated." This option can be ordered for nearly every skylight on the market. It's best to check with a local building official to find out if any stipulations apply to your installation.

Most skylight brochures include performance tables for the different glazing configurations. Those tables are usually broken down into four categories: light transmission, shading coefficient, UV-blockage and U-value. Light transmission is the amount of light allowed through the glass. Shading coefficient is the amount of solar-heat gain through the glass as compared with a single pane of ⅛-in. clear glass. The lower the shading coefficient, the lower the solar-heat gain.

The percentage of the sun's ultraviolet radiation stopped by the glazing is the UV-blockage number. The sun's UV-rays can fade and degrade furniture, carpet and draperies, and they can even discolor wood floors. The last category, U-value, is a measure of heat transfer through a glazing system. The lower the U-value, the better the insulating performance of the glazing system. With seemingly endless combinations and permutations of glass types and coatings, the insulating performance of glass is a topic worthy of a separate article. But I'll try to give a brief description of the options that are available.

maintenance. Motorizing any of Roto's Sunrise II model skylights adds $95 to the price.

The mechanisms that hold crank-out skylights open fall into two categories. One variety has metal arms that swing, or scissor, out to open the skylight. Both Velux and Andersen (100 4th Ave. N., Bayport, Minn. 55003-1096; 800-426-4261) use this type of apparatus, similar to the opening mechanisms for awning and casement windows.

The second and most popular system among skylight companies opens the skylight with a chain that uncoils and stiffens as the crank handle is turned. The chains are made either of metal or of plastic, and manufacturers using this system brag that their skylights open farther than those using the stiff-arm mechanisms.

Like window glass, skylight glass can be coated, such as with a tint or a low-E coating, to affect heat and light transmissions. However, the effectiveness of the various coatings depends on which of the four surfaces in the insulated-glass sandwich are coated.

Southwall Technologies (1029 Corporation Way, Palo Alto, Calif. 94303; 800-365-8794) has put a new spin on glazing performance with its Heat Mirror products, which have a clear film suspended between two sheets of glass. Southwall's Superglass, arguably the most efficient glazing option for skylights, has two film layers between the glass sheets.

The right combination of glass types and coatings will depend a lot on your situation (photo right). For instance, if you live in warmer climate and you want to put a skylight on a south-facing roof, a tint or a coating that cuts down on solar gain may be more important than the insulating value of the glazing. You'll also want a skylight with a high UV-blockage number. One note of caution: If you choose a skylight with tinted glass, the color of the tint can affect the color of everything lighted by the skylight. The added cost of special coatings and glass configurations varies among skylight manufacturers. The difference in price between Velux's FS 106 unit with clear tempered glass and the same skylight with laminated glass with a low-E coating is only $30, which seems like a bargain given the prospective energy savings over the life of the skylight.

Glazing performance is not the same for windows and skylights— A lot has happened recently in the study of glass performance. The National Fenestration Rating Council (1300 Spring St., Suite 120, Silver Spring, Md. 20910; 301-589-6372) certifies the performance of windows, doors and skylights from the major manufacturers with all of their various glazing options. But the numbers provided in the NFRC Certified Products Directory come from testing all of these skylights in a vertical position and may be suspect. When insulated glass is put in a slanted configuration, as with a skylight on a roof, its internal dynamics change dramatically (drawing right). The U-value of certain glazings can degrade up to 30% when changed from vertical to a slope of just 27°.

Skylight glazings are available in a mind-boggling variety of coatings and configurations that affect skylight performance. The NFRC is testing each of these variations and combinations in a sloped orientation and compiling new ratings for each skylight manufacturer. Until those new numbers are out, the NFRC's U-value ratings for skylights should be viewed with some skepticism.

Skylights are designed to accommodate conventional roof framing— Skylights have always been made to fit between regularly spaced rafters, either 2 ft. or 16 in. o. c. If you're looking to put a skylight between rafters 2 ft. apart, you'll find a variety of skylights with a width of 22 in., usually plus a fraction. If your rafters are 16 in. apart, every manufacturer offers a skylight that will span two bays, and many offer skylights that will fit into three 16-in. bays or two 2-ft. bays.

Roof trusses can limit skylight options. In new construction trusses can be engineered to accommodate wider skylights (top photo, p. 92). But in a remodel you've got to deal with what's there, and altering trusses without an engineer's approval is a no-no. If you are one of the unlucky ones who wants to put a skylight in your 1970s raised ranch but the trusses in your roof are 16 in. o. c., take heart. Roto makes the Sweet 16, a fixed skylight that fits into 16-in. o. c. truss or rafter bays. This spring, Roto is introducing a Sweet 16 operating skylight.

Another alternative to cutting rafters is ganging skylights together. Many manufacturers offer flashing kits for side-by-side installations. According to Chuck Silver of Hudson River Design in New Paltz, New York, the best combo kit belongs to Crestline (P. O. Box 8007, Wausau, Wis. 54402-8007; 800-444-1090), which incorporates a raised fin on its skylights to make ganging almost foolproof; they are also good at keeping out the weather.

Skylight glazing should fit the application. Clear tempered glass was chosen for the skylights in this sunroom to let in plant-friendly light. Skylights that open with a latch are less convenient but were chosen because they open wider for maximum ventilation in summer.

Insulated-glass performance changes when installed on a roof

When insulated glass is moved from a vertical to a sloped orientation, the internal dynamics change dramatically. Between the panes of glass, convection currents cause heat to be transferred from the warmer inside surface to the cooler outside.

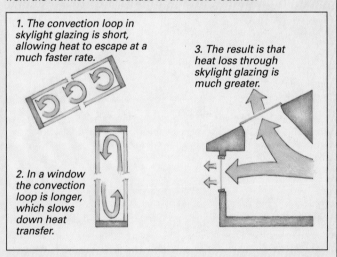

1. The convection loop in skylight glazing is short, allowing heat to escape at a much faster rate.

2. In a window the convection loop is longer, which slows down heat transfer.

3. The result is that heat loss through skylight glazing is much greater.

Flashing is the key to a good installation— I installed my first skylights back in 1982. The guy I was working for handed me two Velux venting skylights still in the box and told me to prep them for installation. I spent a whole morning removing the aluminum cladding and fastening the mounting brackets at just the right height. Luckily, my boss had installed skylights before, and I watched in awe as he wove in the flashing. My duties were to prevent all of the pieces from blowing away and to keep the screws from jumping off the roof until they could be put back in the cladding. After those skylights were installed, they looked as if they could fend off any kind of weather or an attack from a Klingon squadron.

Aside from the glazing choices, the biggest distinction between different makes of skylights is the flashing. In 1941 in Denmark, V. Kann Rasmussen

Wide skylights in truss roofs. A truss roof can be engineered to accept a wide skylight. Here, trusses have been doubled, and a small monotruss that fills in the roof below the skylight hangs on a header.

A shiny alternative finish. Sun-Tek offers its Classic Series skylights in polished copper with a clear, protective finish designed to keep the skylight shiny for years. Photo courtesy of Sun-Tek.

designed one of the first self-contained skylights. His work evolved into what is now the Velux company. Part of his design was the patented Velux flashing system, which today remains basically unchanged (photo bottom right, facing page). One-piece head and sill flashings wrap the top and bottom of the skylight, and step flashing seals the sides. Cladding on the sash, which is removed and replaced during installation, extends down over the top edge of the flashing, serving as counterflashing to create an impenetrable shell.

Other skylight manufacturers such as Roto and Andersen use an EPDM rubber gasket that covers the top of flashing (photo top left, facing page). Mike Guertin, a builder in East Greenwich, R. I., says that he prefers Roto skylights because with the gasket, the sash and all of its cladding do not need to be removed for installation, making the process go a lot quicker.

On the other hand, curb-flashed or perimeter-flashed skylights with a solid flange that runs around the circumference of the skylight always worried me. I'd always associated this type of flashing system with less expensive skylights—that is, until I began my research on this article.

I found many well-made skylights available with welded metal, vinyl or flexible-PVC flanges (photo top center, facing page). The skylights with metal or vinyl have to be installed in mastic to make them waterproof,

and the word *mastic* always conjures up images of goo stuck in my beard and sleepless nights as I worried that the mastic seal might not be complete. Nevertheless, many contractors will use nothing else; in fact, these mastic marvels are popular for commercial installations.

The E-Class skylight made by Wasco has a flexible-PVC flange extruded as an integral part of a PVC frame (photo top right, facing page). The flange has an inverted "L" about an inch away from the frame; the roof shingles slip under the inverted "L." Outboard of the "L" are three water-diversion ridges. The design of the flange eliminates the need for mastic or sealant and is self-healing so that it can be nailed without worry. E-Class skylights are carried to the roof in one piece and installed with small metal brackets that lock into a channel in the skylight frame. Installation of a Wasco skylight takes a fraction of the time it usually takes for a step-flashed skylight.

Sun-Tek (10303 General Drive, Orlando, Fla. 32824; 800-334-5854) offers a variation on the same theme. Its Elite series uses a welded aluminum flange with similar-looking water-diversion ridges. Predrilled, the flange doubles as a nailing flange and eliminates the need for extra brackets. However, installation of the Elite does require mastic or sealant.

Perimeter-flashed skylights also are installed on top of the roof sheathing, which lets the skylight frame be as large as the framed opening in the roof, while the frames for step-flashed skylights usually fit inside of the roof framing. A larger frame means more glass area and more daylight. Wasco's E-Class 2246 venting skylight has 6.39 sq. ft. of glass compared with 3.99 sq. ft. with Velux's VS 106, or 2.4 sq. ft. more daylight for basically the same-size hole in your ceiling. In all fairness to Velux, this spring the company's skylights will be featuring more streamlined and more installer-friendly flashing that will increase glass area for their VS 106 skylight to 4.68 sq. ft.

Special flashing for shallow pitches—The flashing systems I've discussed so far are restricted to installations on roofs with a 4-in-12 pitch or greater. Most skylight companies offer special flashing kits for skylights on shallow pitches, but many of these shallow-slope kits are cumbersome. Because these flashing kits raise the skylight to a higher pitch than the roof, framing and finishing the skylight chase can be a real puzzle. Another alternative for a shallow-pitched roof is a curb-mounted skylight.

Tom O'Brien, a restoration carpenter in Richmond, Virginia, explained that installing a curb-mounted skylight on a shallow roof (4-pitch or less) involved first building a 2x6 frame or curb on top of the sheathing. Tom usually hires an experienced roofer to fabricate metal flashing around the curb. The flashing is embedded with mastic or sealant, and the roofing material is then run on top of the flashing. The curb-mounted skylight sash with built-in counterflashing is then installed on top of the curb in a bed of mastic or sealant. On a flat roof the roofing membrane is carried up the sides of the curb with the sash mounted on top.

Special flashing kits are also needed if your roof is covered with something other than asphalt or wood shingles, such as metal or tile. If it is, be sure the company that makes the skylight you choose also makes the right flashing kit for your type of roof. This type of installation can be tricky, and I also recommend leaving it to a professional roofer.

Step-flashed skylights are harder to make airtight—Although step-flashing is a great system for keeping water out, its weakest point probably is preventing inside air from escaping. No doubt it must seem strange to consider such a factor when venting skylights are designed specifically to allow air to escape, but air leakage around skylights can be a big problem, especially in colder climates.

The problem is not with the step-flashed skylight itself, but rather with the installation. Sealing a step-flashed skylight against air leakage requires felt paper to be run from the roof up the sides of the skylight frame, which means that the installer has to haul yet another item onto the roof in addi-

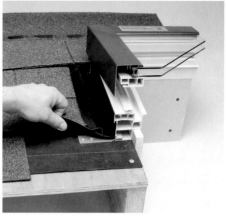

A rubber gasket seals the flashing. Many skylight manufacturers use rubber gaskets to cover the top of the flashing, such as the Roto skylight pictured here.

Curb-flashed skylights rely on a mastic seal. When the mastic seal around these curb-flashed skylights failed, roof cement was incorrectly applied on the shingles to stop a leak.

A new wrinkle in skylight technology. Wasco's E-Class skylight, shown here in cross section, has a PVC frame extruded in one piece with a flexible flange for flashing.

Bare shingles can indicate air leakage. Step-flashed skylights have to be sealed with felt paper under the flashing to make them airtight. If not done properly or skipped altogether, warm air leaking from inside will melt the snow around the perimeter of the skylight.

Tried-and-true flashing system. Velux's patented system consists of head flashing for the top of the skylight, sill flashing for the lower end and step flashing for the sides. Cladding on the sash seals the top of the flashing from the weather. Photo courtesy of Velux-America.

tion to all of the flashing pieces. If this step is skipped or not done properly, air from inside can escape through the spaces between the step flashing, along with lots of heating dollars. After a snowstorm here in Connecticut, it's common to see skylights on snow-covered roofs with halos of bare shingles (photo bottom left). In some cases melting can occur because of heat loss through the skylight frame, but air leakage is usually the culprit.

Skylights with solid flanges don't require felt-paper seals. The folks at Wasco do a little demonstration where they place a $50 bill under one of their fixed E-Class skylights on top of a solid table. Anyone who can lift the skylight off the table can have the bill. But because the flexible flange forms an airtight seal, the skylight won't budge, and no one has won the $50 yet. The flange functions the same way on the roof, forming an airtight seal against the roof sheathing and preventing warm air from finding its way out and around the skylight.

Designer skylights are available in colors—Most companies offer the choice of just a couple of colors. Roto, however, offers its Sunrise II skylights in five different colors including forest green and fire red. Other companies will custom-paint skylights to match or complement any funky color you might have on your roof. A word to the wise: Check out price

and lead time before ordering your sea-foam green skylights. You may decide that basic brown won't look so bad after all.

To me, the neatest-looking skylights are the copper-clad skylights available from Velux and Sun-Tek. Velux's copper-clad skylight is unfinished so that it will weather to a green patina. The Velux copper-clad skylight may be the best choice when skylights are used on historic buildings.

Sun-Tek's copper-clad skylight has a clear, protective finish for a shiny copper-kettle look (bottom photo, facing page). In the right application, this skylight has my vote for sexiest skylight on the market. The cost of upgrading to a copper-clad skylight is reasonable, adding $61 to the price of Velux's VS 106 and $33 to the price of Sun-Tek's VCG 2246.

One final word about availability. Most lumber stores carry many different makes and models and can order what you need if they don't have it in stock. If you are stuck for time and need an out-of-stock skylight yesterday, go with Velux. They pride themselves on being able to ship any of their standard skylights or accessories to your dealer free of charge within 24 hours of receiving an order. ☐

Roe A. Osborn is an associate editor at Fine Homebuilding *magazine. Photos by the author except where noted.*

Frost-Protected Shallow Foundations

Used overseas successfully for decades, this innovative foundation system uses carefully placed insulation and a slab with integral footings instead of conventional deep frost walls

by Christopher R. Kendall

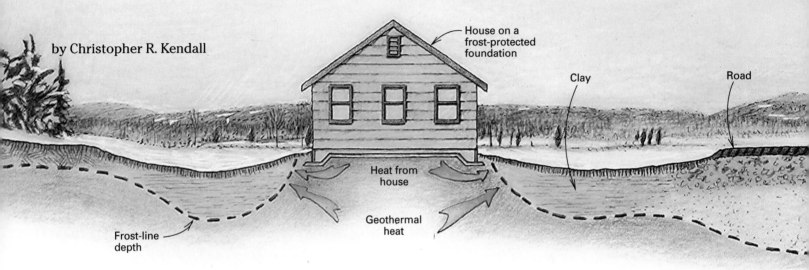

House on a frost-protected foundation

Clay

Road

Heat from house

Geothermal heat

Frost-line depth

I used to work as project manager for a company that builds town houses on top of slab-on-grade foundations. For years we poured our slabs on top of conventional deep foundations built with insulated concrete block. After reading a report from the National Association of Home Builders (NAHB) Research Center (800-638-8556) that is now called *Design Guide for Frost-Protected Shallow Foundations*, I convinced my foreman, Fred Mellott, and the local code-enforcement officer to give this new type of foundation a try. As Fred and I designed the foundations for the 25-unit single-story townhouse project on our new site, the NAHB report, which was funded by a federal grant, became our bible.

Admittedly, our first frost-protected shallow foundation did not go as smoothly as we'd hoped. The learning curve was steep, but as often happens, Fred and the crew helped to figure out the best methods for site preparation and form assembly as the first project progressed. By the time the first building was ready for concrete placement, we were up to speed, and watching the crew put in the foundation for the next unit, you'd have thought they had been building shallow foundations for years. We were convinced. We'd never go back to digging deep

footings, lugging concrete block and dealing with cranky masons, not to mention backfilling and tamping all that soil.

Thawing out conventional frost wisdom— Frost-protected shallow foundations have been standard building practice in Scandinavian countries since World War II. The idea evolved as a sensible alternative to the usual deep foundation set on a footing for colder regions such as northern Europe, the northern United States and Canada. Frost-protected foundations are less costly in time and materials than deep foundations, and they need less site disturbance.

But doesn't a foundation have to be put below the frost line, where it won't be damaged by the freezing and expansion of moisture in the soil, and isn't that line many feet below ground level during the coldest months of the year? To answer these questions, you must first realize that the frost line, or the depth to which frost penetrates the ground, can vary significantly depending on the type of soil and what is on top of that soil (drawing above). For example, the frost line may be relatively shallow in the ground under a wooded area, but a road on a base of sand or gravel allows the frost to penetrate to a much deeper level.

Likewise, the frost line is usually higher around a heated house. Heat escaping from the house into the ground through the slab and foundation plus geothermal heat, or the warmth of the earth trapped beneath the slab, causes the frost line to rise around the perimeter of the foundation. A frost-protected foundation redirects that warmth into the soil by strategically placing insulation on and around the outside of the foundation, thus raising the frost line at the foundation perimeter even more. In fact, with proper insulation most frost-protected shallow foundations need to be only 16 in. deep even in the coldest of climates.

Footings are built into the edges of frost-protected foundations—In addition to the heated living spaces, the town houses we built have attached garages and screened-in porches on slabs (photo facing page). We used frost-protected shallow foundations for each situation, but the exact configuration depended first on whether the space was heated.

If the foundation was going under a heated space, the main area of the slab was excavated down about 8 in. from the projected floor height (top drawing, pp. 96-97). About 2 ft. from the outer edge of the foundation, we sloped the soil

Drawings: Christopher Clapp

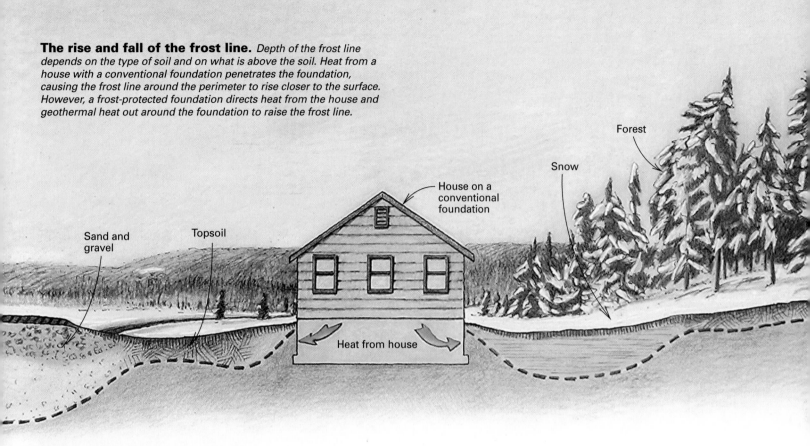

The rise and fall of the frost line. *Depth of the frost line depends on the type of soil and on what is above the soil. Heat from a house with a conventional foundation penetrates the foundation, causing the frost line around the perimeter to rise closer to the surface. However, a frost-protected foundation directs heat from the house and geothermal heat out around the foundation to raise the frost line.*

Forest

Snow

House on a conventional foundation

Topsoil

Sand and gravel

Heat from house

Frost-protected support for any situation. Every section of these town-house units was built on shallow, frost-protected foundations. With properly placed insulation, these foundations can be adapted for heated and unheated spaces as well as for garage stemwalls.

down to a depth of 20 in. at a 45° angle. After the soil was compacted, a 4-in. layer of crushed stone was spread over the entire area, and we were ready for our vapor barrier and forms.

Under an unheated space such as a porch, we took the entire area of the slab down to a depth of 22 in. (bottom drawing). After the soil was compacted, the whole area was covered with a 4-in. layer of crushed stone with 2-in. extruded-polystyrene (EXPS) foam on top of that. The stone and the foam extended beyond the foundation a couple of feet except where the porch abutted the house.

Next, we covered the foam under the main area of the slab with 12 in. of crushed stone. We sloped it down to the level of the foam board around the perimeter to form the same built-up edge that we created under the heated space. If the unheated space was a garage, we formed stemwalls on top of the foam (top photo, p. 99).

We opted for the stemwall rather than a complete slab system for a couple of reasons. First, we are required by law to step the garage floor down 4 in. below the floor of the living space. Putting a monolithic slab at this level would mean not being able to use precut studs, which we were using for the rest of the house. Plus, we liked to pitch the garage floor toward the apron, and pitching the entire foundation structure would make each stud a different length. A stemwall with its top at the same level as the house slab solved both problems.

Easy-to-build forms are reusable—Once the ground had been prepared, we were ready to set our forms. Our most basic form consisted of ½-in. oriented strand board (OSB) ripped to 12 in. and rabbeted into a 2x4 top and bottom (bottom photo, facing page). We covered the crushed stone with a 6-mil poly vapor barrier, then set our form panels on edge, keeping them vertical with 30-in. steel form pins (stakes with holes in them for nails) ¾ in. in dia. We added diagonal oak bracing at every pin to keep the forms from blowing out under the weight of the concrete. If we were forming a foundation to go under a heated space, the inside of the form was lined with 2-in. EXPS, which stuck to the concrete (top drawing).

Here in Pennsylvania, placing the insulation inside the forms provided adequate frost protection, according to the HUD report. In areas farther north, additional insulation called wing insulation must be placed around the perimeter of the foundation to keep frost at bay. The width and the thickness of insulation varies with the severity of local climate; the HUD report offers detailed recommendations for specific areas.

All our town houses have brick-veneer front walls that put a fly in the foundation-design ointment. The problem was how to support brick

Forming a foundation under a heated space
The insulation around the perimeter of a frost-protected foundation under a heated space channels heat from the house into the ground, keeping frost away from the foundation. In colder climates wing insulation, buried in the ground outside the foundation, may be required to trap additional geothermal heat to protect the foundation.

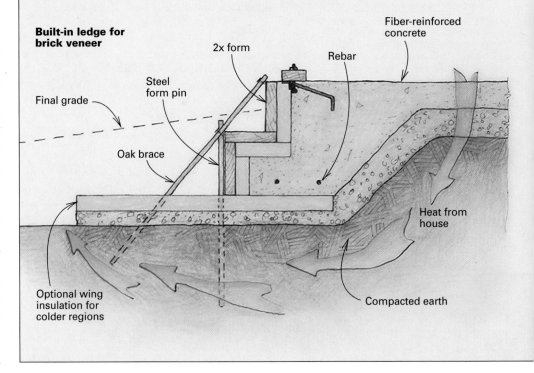

Built-in ledge for brick veneer

Final grade

Steel form pin

2x form

Oak brace

Rebar

Fiber-reinforced concrete

Heat from house

Optional wing insulation for colder regions

Compacted earth

Forming a foundation under an unheated space
Insulation that runs under an unheated space such as a porch traps geothermal heat to protect the foundation from possible damage from frost. Porch slab fits into an integral ledge formed in the edge of the heated space.

Integral ledge between heated and unheated spaces

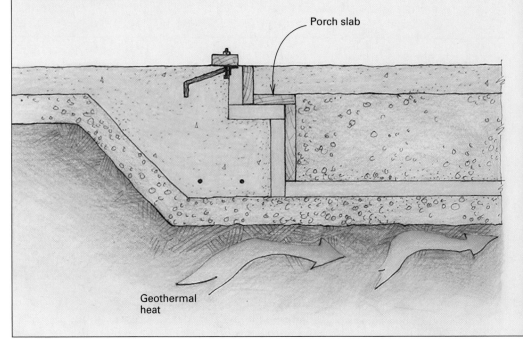

Porch slab

Geothermal heat

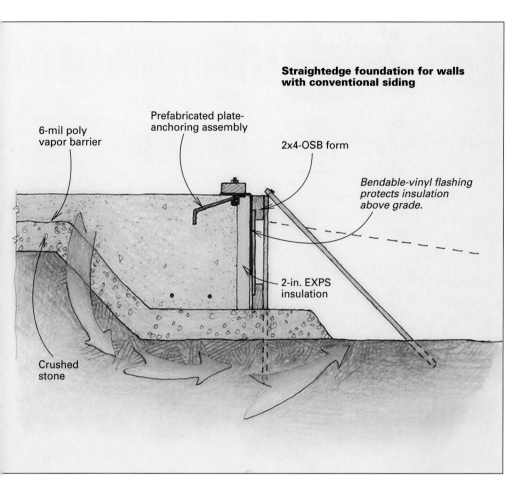

**Straightedge foundation for walls
with conventional siding**

6-mil poly
vapor barrier

Prefabricated plate-
anchoring assembly

2x4-OSB form

*Bendable-vinyl flashing
protects insulation
above grade.*

2-in. EXPS
insulation

Crushed
stone

**Conventional anchor bolts would have
been too near the slab edge.** To anchor
2x4 walls, a plate-support system made out of
steel angle and rebar is prefabricated and installed along the top edge of the foam board
before the concrete is poured.

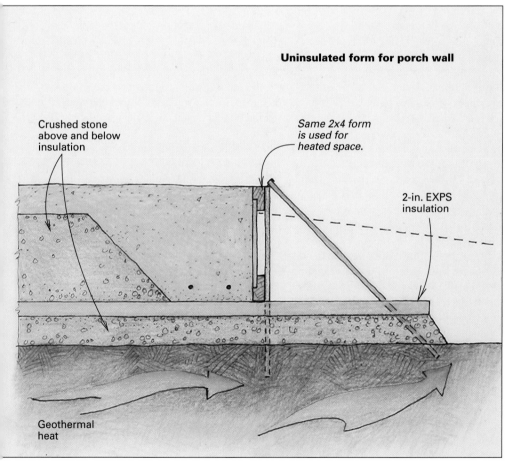

Uninsulated form for porch wall

Crushed stone
above and below
insulation

*Same 2x4 form
is used for
heated space.*

2-in. EXPS
insulation

Geothermal
heat

Site-built forms can be reused. Simple
forms that are made from oriented strand
board and 2x stock are put together on site
and held in place with form pins and wood
bracing. The same forms can be used for each
phase of the pour.

veneer without compromising the rigid insulation and creating a cold bridge. Our bible, the HUD report, never addressed this detail because none of its test houses had brick. Our idea was to support the brick veneer on a layer of foam insulation that would sit on top of a concrete ledge that would be part of the monolithic slab.

The only drawback to this solution was that the Brick Institute of America had deemed EXPS insulation to be inadequate support for brick. But using their figures, we figured that the load exerted by an 8-ft. column of standard brick was about 8 psi. U. C. Industries rates the compressive strength of its Foamular EXPS insulation at 25 psi, so we risked using this unproved, undocumented detail. In the two years since we built our first brick-veneered wall this way, we have experienced no problems with settlement or mortar-joint failure.

We made the form for this detail from 2x8s ripped to the proper width and screwed together in the shape of a step (top drawing, p. 96). This stepped form is lined with foam board and held in place with form pins and diagonal bracing. A similarly shaped form was used where the unheated space meets a heated space. In this case the stepped forms create an integral concrete ledge to support the slab of the unheated space, rather than the brick veneer (bottom drawing, p. 96). The same 2x4-and-OSB form that we used for the foundation under a heated space is also used on the outside edge of an unheated space, except that the forms are on top of the insulation (bottom drawing, p. 97).

We flipped the form for our brick ledge upside down to use as the inside form for our garage stemwalls (drawing right). The inverted form left us enough vertical space on the stemwall to pitch the 6-in. garage slab toward the door and to have concrete on top of the stemwall base.

Shop-welded steel helps to anchor the walls—Just before the concrete truck arrived, we added two courses of steel-reinforcement rod set on chairs that hold the rebar at 3 in. from the bottom of our built-up edge. We also added a special system for anchoring exterior walls.

For our town houses to qualify for an energy grant from the local power company, we needed to build our units with R-18 sidewalls. The easiest way to achieve this R-factor was by using 2x6 construction. However, because of the extra cost in materials and because of our attempt to maximize available living space, we decided on 2x4 walls with ¾-in. polyisocyanurate sheathing to provide the insulation to qualify for the grant.

Unfortunately, our wall plan created a problem with supporting and anchoring the wall plates. To keep the sheathed walls flush with the outside of the 2-in. EXPS foam on the foundation, the 2x4 bottom plate had to cantilever be-

Forming frost-protected garage stemwalls
To maintain consistent wall and floor heights, stemwalls are poured to support the garage. A pitched garage slab is then poured between the stemwalls. As with the unheated space, insulation under the slab forces geothermal heat out to warm the soil around the perimeter of the foundation.

Garage stemwall for brick veneer

The same 2x form for brick ledge is inverted for concrete ledge.

Brick ledge

Concrete ledge

Geothermal heat

yond the edge of the slab 1¼ in., and the remaining 2¼ in. of the plates would not be adequate to carry the wall loads.

We solved the problem with steel. A local welding shop fabricated a reinforced plate-anchoring system out of 1¼-in. by 1¼-in. angle stock ³⁄₁₆ in. thick (top photo, p. 97). The shop welded #4 rebar hooks and ½-in. anchor bolts onto the angle, and the whole system cost only $2 per ft. The angle stock was rabbeted into the inside edge of the foam board, and angle irons slipped over the anchor bolts and extended over the edge of the forms to hold the whole system in place during the pour.

A vibrator helps to fill the forms with concrete—As with any slab, the locations of plumbing stubs and any other under-the-floor systems were double-checked according to the floor plan before we ordered the concrete. When everything was set, we began by pouring all but the top 4 in. in each section (bottom photo, facing page). We specified a 3,500-psi mix with a 4-in. to 5-in. slump, a fairly stiff mix.

Consolidating the concrete beside our straight vertical forms was easy. But the stepped forms that create the ledges for the brick veneer and the integral concrete slab presented a bigger challenge. It was imperative to fill the space be-

neath the horizontal sections of the form completely with concrete, and a concrete vibrator helped us to ensure complete consolidation in these areas. The head of the vibrator is on the end of a long tube inserted into the wet concrete under the ledge form. As the head vibrates, it forces out trapped air, allowing total consolidation in these areas.

When the edges had been poured and consolidated, we poured the rest of the slab. Wire mesh is ordinarily used to reinforce a slab. However, we found it a headache to get the mesh at the proper height or to pull it through the wet concrete and have it stay while the crew slogged around on top of the mesh. Instead, we used concrete reinforced with polypropylene fibers.

Screeding the concrete was another area where we found shallow foundations superior to block foundations. It's not that easy to get a smooth, even surface on the slab when screeding to the irregularities of the top of a block wall. But screeding to the angle stock of our plate-anchoring system was quick and easy.

While the concrete was still green, we used a diamond-blade saw to cut contraction joints in the slab. Contraction joints minimize the random cracking that occurs naturally as cement cures. With planning, most of these sawcuts can be hidden under interior partitions.

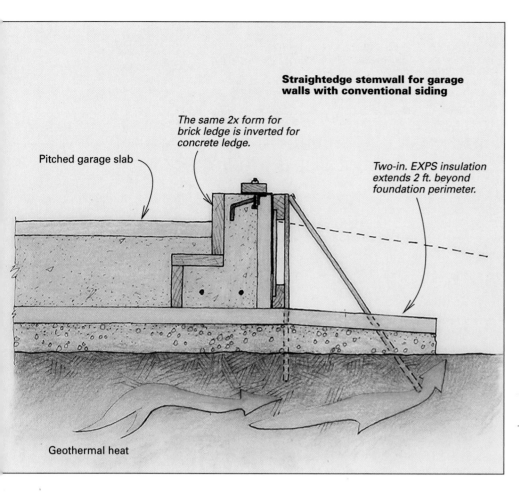

Pitched garage slab

Straightedge stemwall for garage walls with conventional siding

The same 2x form for brick ledge is inverted for concrete ledge.

Two-in. EXPS insulation extends 2 ft. beyond foundation perimeter.

Geothermal heat

Insulation extends beyond the stemwalls. Garage stemwalls are formed directly on top of the foam-board insulation. The insulation continuing beyond the stemwall perimeter traps heat from the earth to keep frost at bay.

Bendable vinyl keeps out the bugs and the UV-rays—When test houses were built in this country using frost-protected shallow foundations, one problem builders faced was covering the exposed EXPS foam insulation between the grade line and the siding starter course. EXPS degrades when exposed to ultraviolet (UV) radiation and doesn't offer much resistance to impacts from baseballs and string trimmers.

Builders of test houses tried parging, brush-applied coatings and treated plywood. But we came up with a different solution that's simple and easily installed. Bendable-vinyl coil stock manufactured by Pro-Trim (Alum-A-Pole, 1011 Capouse Ave., Scranton, Penn. 18509; 800-421-2586) comes in 24-in. wide by 50-ft. long coils and is perfect for this application.

We wrapped the top outer edge of the foundation beginning at the inside edge of the plate. The vinyl bends easily with a standard siding brake to wrap over the foam board and neatly finishes the gap between grade and siding. We glued it to the edge of the slab and to the EXPS with foam-compatible mastic.

Bendable vinyl glued in place also provides an insect barrier. Insect penetration is an area of some concern because certain insects seem to excel at tunneling through foam board. So far, they haven't been able to get by the vinyl and

mastic. We filled gaps between the foam board with Great Stuff expanding foam (Insta-Foam Products, 1500 Cedarwood Drive, Joliet, Ill. 60435; 800-800-3626).

Saving money for the contractor and owner—All in all, I figure the cost savings of building frost-protected foundations over conventional methods at just over 3% of the total hard cost per unit. The beauty of frost-protected foundations is that nothing is sacrificed with this method. Although we built in the Northeast, these construction techniques can be applied to areas of high wind-loading or seismic activity.

Another benefit of this type of construction is the ground-coupling effect created by insulating the perimeter of the structure. The large, enclosed thermal mass helps to prevent noticeable temperature swings in living spaces. We installed high-performance air-to-air heat pumps, and with electricity at about 5¢ per kwh, heating during an average season costs less than $300. Likewise, air-conditioning costs average less than $100 for a normal summer. These savings kept the client happy for years to come. □

Christopher R. Kendall is the former project manager for S & A Custom Built Homes Inc. in Chambersburg, Pennsylvania. Photos by the author.

The concrete at the perimeter is poured first. The thicker edges of a frost-protected foundation are poured and consolidated before the top layer is poured for the main slab. When the final layer is poured, the plate-support system acts as a screed guide.

Wall-Sheathing Choices

Today's sheathing materials offer strength, energy-efficiency, economy, and fire-and water-resistance—but not all in the same package

by Bruce Greenlaw

Until the end of World War II, most wood-frame houses in the United States were sheathed with sawn boards. Builders debated whether the boards should be installed horizontally or diagonally. Diagonal sheathing eliminated the need for additional wall bracing, but shrinkage sometimes distorted building frames, causing problems such as cracked stucco. Horizontal sheathing didn't brace walls, but it didn't distort them either and was easier and more economical to install. Unresolved, the issue became moot when plywood superseded board sheathing shortly after World War II.

Nowadays, most houses are sheathed with panels instead of boards (photo below). There are at least a dozen types of wall-sheathing panels on the market, most of which are structural panels that can brace walls or insulating panels that can't brace walls. Fire-resistant panels are also available, some of which can brace walls but none of which insulate. And to top this off, only certain types of structural sheathing can serve as nail base for siding materials that need intermediate fastening between studs, such as wood shingles.

When choosing sheathing materials, you need to consider not only structural, energy and combustion issues, but also code-compliance, cost, availability, job-site durability, ease of installation and, believe it or not, curb appeal (see

OSB sheathing is suitable for structural use. Both plywood and OSB panels are strong enough to brace walls against racking and in most circumstances can be used interchangeably. Sheathing walls before tilting them up can speed framing and simplify the cutting of window openings.

chart p. 102 for a side-by-side comparison of the various sheathing types).

Structural sheathing is designed to brace walls—Although plywood was available as early as 1905, it wasn't until 1938 when waterproof glue replaced hide glue that plywood caught on as a building material. By the postwar building boom of the late 1940s, plywood was well-established as high-quality structural sheathing.

Oriented strand board (OSB), on the other hand, came to market in 1981. Whereas plywood veneers usually are peeled from large-diameter logs, OSB has cross-laminated layers of compressed wood strands machined from small-diameter, fast-growing trees. OSB is nearly as strong as plywood and costs less, which is why it outsells plywood wall sheathing in many areas.

Structural-wood panels such as plywood and OSB are regulated by an approved certification agency (usually APA-The Engineered Wood Association) to verify that the panels meet code-recognized performance standards (top photo). Trademark/grade stamps certify a panel's compliance and list pertinent information about a panel's thickness, exposure durability and structural rating.

Plywood and OSB wall sheathing are normally $\frac{5}{16}$ in. to $\frac{3}{4}$ in. thick, 4 ft. wide and 8 ft. to 10 ft. long. Building codes permit the use of $\frac{5}{16}$-in. panels over studs spaced 16 in. o. c. and $\frac{3}{8}$-in. panels over studs spaced 24 in. o. c., but this can make a flimsy backing for siding. Most builders opt for $\frac{3}{8}$-in. panels over 16-in. framing and either $\frac{7}{16}$-in. or $\frac{1}{2}$-in. panels over 24-in. framing. If shakes or shingles will go on top, use sheathing at least $\frac{5}{8}$ in. thick so that shingle fasteners hold fast.

Structural panels can be installed vertically, though installing them horizontally provides more stiffness for a given panel thickness. The APA recommends horizontal application to walls that will be stuccoed, but this can also make sense when the sheathing will back wood shingles. Horizontal edges between panels normally need to be blocked, which is why most builders prefer to install the panels vertically to save on labor. If you do install panels horizontally, stagger the vertical joints to distribute fasteners evenly throughout the wall.

Regardless of panel orientation, space panels $\frac{1}{8}$ in. apart (the diameter of an 8d nail) at the edges to allow for moisture-induced expansion. Although this practice is often ignored in the field, I've seen tight joints buckle, leaving a raised ridge. Panels that are "sized for spacing" (indicated on the trademark) are undersize lengthwise and widthwise by up to $\frac{1}{8}$ in., which makes it easy to space panels while keeping the vertical joints centered on studs.

Questions still remain about the ability of OSB to resist impact and blow-off during hurricanes.

If you live in a hurricane zone, check with your building department to make sure it's okay to use OSB. Or use plywood.

When shopping for standard plywood wall sheathing, ask for CDX plywood ("C" and "D" refer to the grade of veneer used on the sides of the panel, and "X" refers to the use of exterior glue), which is the trade term for Exposure 1, performance-rated sheathing. Exposure 1 panels can withstand temporary exposure to weather during construction. Structural 1 panels have about 10% more shear strength than standard panels do and commonly are required for sheathing shear walls (stiff walls that are anchored to foundations to resist the high lateral loads typically generated by earthquakes and high winds).

Gypsum sheathing is fire-resistant—Gypsum sheathing (bottom photo) is a common choice for commercial work but accounts for only about 5% of the residential wall-sheathing market. Still, it can make a good substrate for many residential-siding materials, though it typically

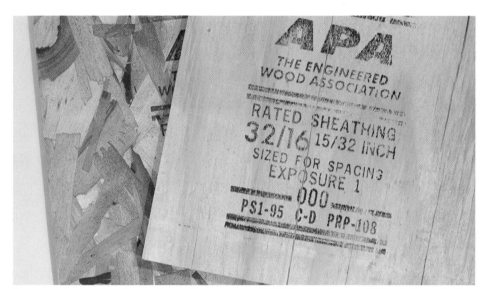

Approved structural-wood panels carry an APA-The Engineered Wood Association certification. Besides certifying that these panels conform to code-approved specifications, the APA grading stamp also indicates the thickness of the panel, its suitability for exposure to weather and its span rating.

Gypsum sheathing is typically used beneath stucco. Fire-resistance and low cost are the chief selling points of gypsum sheathing. Bottom to top: National Gypsum Company Gold Bond Jumbo square edge, Louisiana-Pacific FiberBond and Georgia-Pacific Dens-Glass Gold.

Wall-sheathing comparison

Sheathing	Advantages	Disadvantages
Plywood	Can brace walls; highest shear strength and impact-resistance; can provide a nail base for wood shingles.	Most-expensive structural sheathing; low R-value.
Oriented strand board (OSB)	Can brace walls; cheaper than plywood; can provide a nail base for wood shingles.	Use might be restricted in hurricane areas; low R-value.
Gypsum	Excellent fire-resistance; can brace walls in some areas; low cost	Low impact-resistance; normally must be covered within 30 days; low R-value; can't be used as nail base.
Cement board	Excellent fire-resistance; can be used as a drainage board for EIFS; rot-proof.	Expensive; doesn't brace walls; heavy; low R-value; can't be used as nail base.
Laminated fiber	Can brace walls; low cost; high resistance to air infiltration; lightweight.	Can't be used as nail base; low R-value.
Fiberboard	Can brace walls; best insulating structural sheathing; low cost.	Most types can't be used as nail base.
Expanded polystyrene (EPS)	Highest R-value per dollar; insect-resistant panels available; lightweight.	Doesn't brace walls; thermal barrier required between sheathing and living space; low compressive strength; can't be used as nail base.
Extruded polystyrene (XPX)	High R-value; virtually water-proof; strongest foam in compression; lightweight.	Doesn't brace walls; thermal barrier required between sheathing and living space; can't be used as nail base.
Polyisocyanurate	Highest R-value per inch; more fire-resistant than other foams; lightweight.	Doesn't brace walls; thermal barrier usually required between sheathing and living space; can't be used as nail base.

isn't structural. It's most commonly used as an underlayment for stucco, but I've seen it installed beneath everything from cedar shingles to limestone veneer.

Gypsum sheathing can be scored with a utility knife and snapped apart just like interior drywall. But unlike standard interior drywall, which would disintegrate quickly outdoors, gypsum sheathing has water-repellent paper faces, and it can also have water-resistant cores for extra protection in moist climates or for use beneath synthetic stucco (EIFS). Manufacturers claim that you don't have to install building paper over gypsum sheathing unless it's required by code or unless the sheathing will be exposed to the elements for longer than a month. I'd consider papering it (and most other sheathing materials) anyway to tie into flashing (see *FHB* #100, pp. 58-63).

Gypsum sheathing's biggest selling point is that it's inherently fire-resistant because gypsum contains chemically bound water. Type-X gypsum sheathing is thicker than the standard variety and has special additives in the core for even better fire protection. Type-X sheathing is a popular (and sometimes code required) choice for fire-rated walls in areas threatened by wildfires or on homes that have less than the minimum required setback from property lines. Standard gypsum panels are either $\frac{1}{10}$ in. or $\frac{1}{2}$ in. thick. Type-X panels are $\frac{5}{8}$ in. thick. Both types come in 2-ft. and 4-ft. widths. The 2-ft. wide panels have V-shaped tongue-and-groove long edges and are installed horizontally with the grooves facing down to shed water. These panels are easy to handle, and rows can be installed in concert with strips of asphalt-impregnated building paper or stucco lath as installers make their way up ladders or scaffolding. The narrow panels provide limited shear strength, though, so they can't be used as wall bracing. Most areas allow $\frac{1}{2}$-in. or thicker 4-ft. wide gypsum panels to brace walls, provided the panels are applied vertically according to code. The next edition of the Uniform Building Code will no longer allow this, though. Four-ft. wide panels come in lengths up to 10 ft. for covering tall walls or for overlapping floor framing.

Two paperless gypsum products are also worth a mention; both are stronger and more weather-resistant than conventional gypsum sheathing. Louisiana-Pacific's FiberBond panels (800-411-2500) have fiber-reinforced gypsum and perlite cores that are sandwiched between two beefy layers of fiber-reinforced gypsum. The fiber, which comes from recycled newsprint, makes these panels the strongest gypsum panels on the market. In fact, the structural panels are the only APA-rated gypsum panels available and, like APA-rated plywood and OSB, bear the APA trademark. FiberBond panels also can be exposed to weather for up to 60 days instead of the usual 30 days.

Georgia-Pacific's Dens-Glass Gold panels (800-225-6119) have a silicon-treated gypsum core surfaced on both sides by fiberglass mats. They also have on one side an alkali-resistant coating that eliminates the need to apply a primer/sealer when installing high-alkaline, adhesively applied EIFS.

Dens-Glass Gold sheathing not only is stronger than normal gypsum sheathing, but it's also the most weather-resistant gypsum sheathing around; you can expose it to the elements for up to six months with no problem. This sheathing might be overkill for most residential work, but I would consider using it where long construction delays or high humidity is common. FiberBond and Dens-Glass Gold sheathing both come in regular or type-X panels that can be installed vertically or horizontally without sacrificing wall strength.

All gypsum panels share one drawback: They can't be used as a nail base for siding because they offer little resistance to nail withdrawal. If you plan to shingle over them, you'll have to add a nail-base sheathing or horizontal wood strapping on top.

Some sheathings are better for specific kinds of siding—Cement board (top photo, facing page) isn't a common residential-wall sheathing, but it has made inroads as a sturdy, rot-proof and fire-resistant underlayment for synthetic stucco, exterior tile and thin-brick veneer. Cement board typically comes in $\frac{1}{2}$-in. thick nonstructural 3-ft. or 4-ft. wide panels consisting of aggregated portland cement with fiberglass mesh embedded in both sides. Joints between panels are reinforced if necessary with exterior fiberglass-mesh tape covered with a special joint compound. Cement board is quite porous, so weather-resistant building paper is required over or under it, depending on the application.

Thin panels can brace walls—Amazing as it sounds, nominal $\frac{1}{8}$-in. thick sheathing panels can serve as structural wall bracing. Laminated-fiber panels, including Energy Brace (Fiber-Lam Inc., P. O. Box 2002, Doswell, Va. 23047; 804-876-3135) panels (bottom photo, facing page) and Thermo-ply (Simplex Products Division, P. O. Box 10, Adrian, Mich. 49221-0010; 517-263-8881) consist of layers of cellulose fiber that are pressure-laminated using a water-resistant glue. Thermo-ply is made primarily of preconsumer and postconsumer recycled material, and Energy Brace uses mostly virgin stock. Panels are treated for water-resistance and usually are faced with polyethylene on both sides, aluminum foil on both sides, or poly on one side and aluminum on the other.

Thermo-ply and Energy Brace panels come in three color-coded grades. Structural-grade panels, which have red lettering on them, span studs spaced 16 in. o. c. Slightly thicker panels with blue lettering on them span studs at 24 in. o. c., or they can span studs at 16 in. o. c. for added stiffness. Standard-grade panels (green lettering) are thinner than the others and are the only ones that can't be used as wall bracing. In some areas, Simplex also sells a black-lettered panel that is slightly thinner than red panels and can also brace 16-in. o. c. stud walls.

Foil-faced panels are sometimes installed next to an airspace to add a bit of R-value to walls or to serve as radiant barriers. The foil also makes a house look like it's gift-wrapped, at least until it gets covered by siding. Nevertheless, if you don't need foil, you might want to stick with polyethylene-faced panels so that siding installers (and passersby) don't get blinded by the glare that foil-faced panels can generate.

Simplex and FiberLam also promote their sheathing as air barriers. Both sell 48¾-in. wide by 8-ft. long panels that can be lapped at vertical seams for extra airtightness (center photo). Four-ft. wide panels are also available. Manufacturers claim that the panels are easy to cut with a utility knife. This is true if you have arms like Popeye's and an unlimited supply of sharp blades. I'd use a circular saw.

Fiberboard panels can insulate and brace a wall—The first fiberboard plant in the United States was built in 1908, the same year people started driving Model T Fords. Developed as a paper byproduct, fiberboard was hailed as one of the first insulating sheathings (bottom photo, p. 104). Since energy has become expensive, though, fiberboard has lost some of its luster. That's because the R-value of the thickest fiberboard panels on the market is just 2.06, as opposed to a maximum of about R-8 for 1-in. thick polyisocyanurate foam sheathing.

Nevertheless, fiberboard sheathing is still readily available in many areas, and for good reason. It repels water, it's economical, and structurally rated panels can eliminate the need for additional wall bracing. It can also make a good sound barrier in noisy neighborhoods, and building scientists seem to like it (along with gypsum sheathing) because it breathes better than most other sheathing materials. This means

Nonstructural cement board is a good choice for thin-brick veneer, tile and synthetic stucco. Because it is porous, cement board, such as USG's Durock sheathing, requires a weather-resistant building paper over or under it to protect the framing. This sheathing can be scored with a carbide knife and snapped apart by hand.

that it allows water vapor to escape from buildings in cold climates instead of trapping it inside walls, where it can condense and cause mildew, rot or other problems.

Many of the fiberboard-sheathing panels on the market are made by members of the American Fiberboard Association (AFA, 1210 W. Northwest Highway, Palatine, Ill. 60067; 847-934-8394). These panels are made of wood and agricultural byproducts (such as sugar cane); are impregnated and coated on at least one side with asphalt for strength and weather-resistance; measure 4 ft. wide by either 8 ft. or 9 ft. long; and come in two types, regular and structural (formerly called "intermediate"). Regular panels are ½ in. thick, and structural panels are either ½ in. or 25⁄32 in. thick.

If you plan to use fiberboard panels as wall bracing, choose structural panels, which are labeled "structural" on the face, and apply them according to code. Regular panels can be used as economical fillers between strategically placed structural fiberboard, plywood or OSB panels. Like plywood and OSB, fiberboard panels should be spaced ⅛ in. apart to allow for expansion, and all horizontal edges should be supported by framing or blocking.

The Homasote Company (800-257-9491), which isn't an AFA member, makes high-density fiberboard sheathing panels out of recycled newsprint. Homasote claims that the panels can be used as a nail base for shingles (ordinary fiberboard can't be), provided the shingles are secured with ring-shank nails. The panels aren't treated with asphalt, so they don't have the char-

Laminated-fiber sheathing has a thin profile, yet is structurally rated. Lightweight ⅛-in. thick structural-grade pressure-laminated cellulose panels by Simplex are poly-faced (left) or foil-faced (right) and are strong enough to brace walls.

Thin sheathing can be lapped at the edges to reduce air infiltration. Laminated-fiber sheathing is available in 48¾-in. wide panels that can be lapped at the seams, which increases its performance as an air barrier.

Foam insulates walls. Left to right: AFM Perform Guard insect-resistant EPS; Dow Duramate film-faced XPS; Dow Bluecor P/P XPS fanfold; Celo- tex Tuff-R foil-faced polyisocyanurate; Celotex Sturdy-R fiberglass-faced polyisocyanurate; Celotex Quik-R polyisocyanurate for EIFS.

acteristic black or dark-brown color of most fiberboard panels, and they aren't as weather-proof. Water-resistant building paper should be installed on top.

Rigid-foam sheathing can insulate a wall but not brace it—Some manufacturers tout their rigid-foam board as an energy-efficient re-placement for other types of sheathing. That's like touting Hondas as energy-efficient replace-ments for pickup trucks: Rigid foam simply does not qualify as wall bracing. Foam also can't serve as a nail base for shingles as some sheath-ings can. And unlike gypsum sheathing, foam can be more of a fire hazard than a fire retar-dant without a proper thermal barrier, though, for instance, Celotex chemically modifies its Thermax sheathing for fire-resistance.

But foam sheathing allows builders to produce the R-19 walls required by code in many areas without switching from 2x4 to 2x6 framing. And because foam envelops wall framing, it reduces thermal bridging through the framing itself and can prevent condensation from forming in the wall cavities.

Most rigid foam comes in 4-ft. wide by 8-ft. or longer panels, but 16-in. wide and 24-in. wide panels are also available for horizontal applica-tion (2-ft. by 4-ft. panels are available in some areas for stucco contractors). The choice of foam-panel thicknesses can be daunting, but the most popular panels range up to 1 in. thick. Some foam panels have tongue-and-groove or shiplap edges for a more airtight installation; others have square edges. Also available are fan-fold panels that are installed directly over old

Fiberboard sheathing can both insulate and brace walls. Although fiberboard's R-value is relatively low, structural fiberboard (bottom) can be used with economical asphalt-coated regular fiberboard (top) to build an in-sulated, breathable and sound-deadened wall.

siding as an underlayment for new siding. These panels are ⅜ in. or ½ in. thick and come in 4-ft. wide by 50-ft. long sheets that are folded like ac-cordions with creases spaced 2 ft. apart.

EPS is a good choice for EIFS siding—Foam sheathing is available in three types: expanded polystyrene (EPS), extruded polystyrene (XPS) and polyisocyanurate (top photo). EPS, or "beadboard," is widely regarded as the environ-mentally friendly foam because it's the only one

that doesn't contain ozone-damaging blowing agents. It's made of steam-puffed polystyrene beads that are molded into blocks and sliced into panels. EPS has the lowest R-value per inch of the three types, averaging about R-4, but it can deliver the most R-value per dollar.

Unfaced EPS sheathing is commonly used in residential EIFS and one-coat stucco systems. It's inexpensive, high spots can be rasped off easily, and panels can be cut quickly and clean-ly into stucco moldings using a hot wire or a hot knife (available through EIFS suppliers). EPS panels faced with polyethylene or foil are also available. Both facers improve weather resis-tance and durability, and foil adds R-value when facing an airspace. Panels treated with nontoxic borate resist termites, carpenter ants and other wood-boring insects.

XPS stands up to the weather—Extruded polystyrene, or XPS, is made of the same plastic as EPS, but XPS is inflated with blowing agents that give it higher R-values per inch (typically R-5). XPS panels come with or without plastic film facers. Most builders prefer faced panels because they withstand rougher handling, but unfaced XPS panels are becoming increasingly popular as an underlayment for stucco.

XPS panels resist blow-off better than the other rigid foams. And because the foam is highly wa-ter-resistant, panels can be exposed to rain dur-ing construction without the worry of water ab-sorption (after all, XPS buoyancy billets are used for floating-marina docks). And XPS panels can eliminate the need for housewrap or other air barriers, especially when the panels are installed vertically with the edges supported by framing. Sealing panel seams with sheathing tape can further decrease air infiltration.

Because some solvents dissolve polystyrene panels, allow siding that's pretreated with water repellents or wood preservatives on the back to dry before it contacts the foam. Also, be careful not to get solvent-based insecticides on polystyrene panels.

Polyisocyanurate panels have the highest R-value per inch—Unlike the polystyrenes, polyisocyanurate is unaffected by solvents. Polyiso is also the R-value king. Panels average R-6.5 per in., but I've seen panels that rate as high as about R-8 per in.

Polyiso panels are faced either with aluminum foil for maximum energy-efficiency or with coated fiberglass for maximum strength. Foil prevents the rapid intrusion of nitrogen and oxygen into the foam, which allows the R-value of the foam to stabilize at the highest level. Placing reflective foil next to a ³⁄₄-in. airspace in a wall increases the R-value of the wall by 2.77. Foil also repels water and prevents the intrusion of water vapor, both of which could reduce the R-value of polyiso foam.

Reflective foil has been accused of overheating vinyl siding on sunny days and causing it to "oil can," or form ripples. But several vinyl-siding manufacturers have told me that reflective foil has a minimal effect on siding temperatures and that oil canning is caused by improper siding installation (driving nails too tight). Regardless, nonreflective, coated-foil facers are available for those builders who want foil without the glare or who don't trust reflective foil next to a particular siding.

Panels with water-resistant, coated-fiberglass facers for added strength and for puncture-resistance are also available. Like fiberglass batts, though, the facers can make your skin itch as you handle the panels, and the panels lose about 25% of their R-value to the facing.

Specially faced polyiso panels, including Stucco-Shield II (Atlas Roofing Corp., 1775 The Exchange, Suite 160, Atlanta, Ga. 30339; 770-933-4461) and Quik-R Wall Insulation (The Celotex Corp, 4010 Boy Scout Blvd., Tampa, Fla. 33607; 813-873-4000), are designed for use as EIFS substrates. These panels don't always need rigid backing as EPS panels do, and they're manufactured with tight thickness tolerances that are supposed to speed troweling and minimize stucco waste and stress cracks. Polyiso doesn't melt, though, so it can't be cut with hot wires or hot knives as the polystyrenes can. It's typically cut with a utility knife instead. □

Bruce Greenlaw is a contributing editor of Fine Homebuilding. *Photos by the author.*

Sheathing-installation tips

• **Use the right fasteners and follow recommended fastening schedules—** All model building codes include fastening schedules for structural-sheathing materials used as wall bracing. The schedules list acceptable fasteners and tell how far apart to space them along panel edges and over intermediate supports.

Plywood sheathing, for instance, is normally applied with 6d or 8d nails spaced 6 in. apart along the edges and 12 in. apart over intermediate studs. But I've had to space the nails closer in seismic areas because of variations in local codes.

Rigid-foam sheathing doesn't brace walls, so there are no code-enforced fastening schedules for it. These schedules are normally supplied by panel manufacturers and are designed to ensure that panels are pulled tight against the framing for maximum airtightness and don't blow off before they're covered with siding. Foam panels that are included in synthetic-stucco systems actually support the stucco and should be installed according to stucco manufacturers' recommendations.

In general, nail-base sheathings such as plywood and OSB are secured with common or deformed-shank nails. Gypsum, fiberboard and the rest of the sheathing panels are more often fastened with galvanized roofing nails.

Gypsum sheathing is sometimes secured to wood framing with type-W screws. Where I live, you can't use screws if the sheathing will brace walls. Cement board can be fastened with special noncorrosive screws because it doesn't brace walls.

To save time, most pros install sheathing with pneumatic nailers or staplers at every opportunity. Nailers, preferably the type that drive round-head nails, can be used for installing structural wood panels. Pneumatically driven staples with ⁷⁄₁₆-in. crowns can secure structural-wood, gypsum, laminated-fiber and fiberboard panels. Wider crowns usually are preferred for foam sheathing.

• **Don't overdrive sheathing fasteners—**Sheathing fasteners should be placed at least ³⁄₈ in. away from panel edges to prevent tearout. Also, be careful not to overdrive fasteners; if they break through the surface of the sheathing, their holding power will drop substantially.

To prevent overdriving of pneumatic fasteners, you might have to adjust the gun to leave fastener heads slightly proud of the sheathing surface, and then hammer them flush. Hand-nailing also helps to suck up panels tight against framing to prevent gaps and to minimize air infiltration.

• **Allow for framing shrinkage—**Solid wood shrinks across the grain as it dries. Multistory buildings framed with solid lumber can shrink considerably between floors, where wall plates and rim joists form a thick sandwich of horizontal framing members. Some builders space wall sheathing ½ in. between floors to allow for this shrinkage. I'd space them ³⁄₄ in. apart, especially with deep joists.

The easiest way to provide this gap is to tack a ³⁄₄-in. wide wood strip over bottom panels to provide temporary support as top panels are fastened to the framing. Joints between rigid-foam panels can be filled with a compressible-foam backer rod and a compatible sealant to prevent heat loss and, in the case of EIFS stucco, moisture infiltration.

Framing shrinkage can be cut in half by using kiln-dried framing lumber, or it virtually can be eliminated by using engineered framing lumber.

• **Sheath first, then cut the openings—**The speediest way to sheath walls is to panel over window and door openings, and then cut out the openings. Openings in walls that are presheathed with wood-base panels can be marked on the panels as they're installed and cut with a circular saw before the walls are raised. If the sheathing is installed on upright walls, openings can be cut from inside the house with a reciprocating saw or with a chainsaw.

Gypsum sheathing can be cut out fast with a drywall cutout tool, which resembles a laminate trimmer, but this can be done only from the sheathed side of the wall. Unfaced or poly-faced polystyrene-foam panels can be cut with a hot knife (available from EIFS suppliers). These and other rigid-foam panels also can be cut with a drywall handsaw, a utility knife or a sharp kitchen knife that's long enough to use the framing as a cutting guide.

Cement board can be cut with a reciprocating saw fitted with a carbide-tipped blade, but most installers score and snap the panels to fit around openings before the panels are installed because it's faster that way and because it doesn't make dust.—*B. G.*

Making Storm Windows

Wood storm windows look better and seal tighter than mass-produced aluminum and vinyl units

by Rex Alexander

There's no doubt about it. Modern aluminum and vinyl storm windows on an old house are ugly. They don't look right. They're not appropriate. And often, they don't even work right. So when David and Lulu needed storm windows for the old house they remodeled, they asked me to give them a bid for custom-made wood storm windows.

Despite the cost—my bid for 58 storm windows was double that of aluminum and vinyl—they went with wood. Wood storm windows help to preserve the beauty of traditional windows, especially for an older home. Also, because they are custom fit and full size, they eliminate drafts.

First, I had a few decisions to make—Before I proceeded with a prototype for the project, I had to make a few choices. What kind of wood should I use? Should I go to the expense of buying a sash cutter for my shaper or just use a rabbeting bit in my router to rout a place for the glazing? What kind of joints should hold the storm windows together? Finally, what kind of weatherstripping should I use inside the finished storm windows?

For the stock, I decided to use kiln-dried yellow poplar. It machines well with little tearout, paints beautifully and is fairly stable. Poplar is also lightweight, which is a real consideration because these windows will be hoisted in or out twice a year.

After doing some research (looking at old storm windows), I decided to order a sash cutter that cuts an ogee profile and the rabbet for the glass all at once. I bought the reversible window sash cutter for my shaper from MLCS (800-533-9298), which also makes the same cutter for a router. Both are $100. By using the reversible sash cutter, I could shape the inside edges of the stiles and rails, then remove the rabbeting cutter, raise the sash cutter and shape (or cope) the ends of the rails. When the male-female profiles were joined, the two would fit snugly.

As for the type of joint, I could have used mortise-and-tenon joints, but to keep the storm windows flush with the outside trim on the house, the storm windows had to be ⅞ in. thick, which wouldn't have left enough meat on each side of the tenons for my taste. I also could have used biscuits, but because I run probably one of the only shops in the country without a biscuit joiner, I eliminated that alternative. I could have used galvanized drywall screws, but I wanted the finished windows to show quality and craftsmanship.

I decided to use doweled joints. I'm comfortable with doweled joinery. And doweled joints are strong. Although the storm windows wouldn't see a

The doweling jig aligns the holes. After marking the dowel locations on the end of this rail, the author clamps the Dowl-It jig onto the workpiece, chucks a homemade stop on his brad-point bit and drills.

Coping rail ends. With the workpiece clamped flat against the shaper table, rabbet and ogee profiles are cut into the end of this rail. The pencil marks on the end of the workpiece show where the dowel holes were drilled.

Custom-made to be durable and tight fitting. *Made of ⅞-in. thick yellow poplar, the storm window is shaped with an ogee profile on the interior side of the glass openings and with a rabbet on the outside to hold the glazing. The middle rail matches the point on the window where the double-hung sash overlaps. The bottom rail contains four vent holes, covered by a pivoting wood cover that seats tightly into a compound angle cut.*

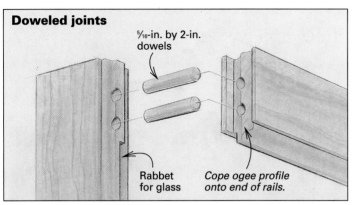

Doweled joints

⁵⁄₁₆-in. by 2-in. dowels

Rabbet for glass

Cope ogee profile onto end of rails.

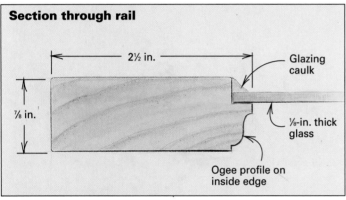

Section through rail

2½ in.

⅞ in.

Glazing caulk

⅛-in. thick glass

Ogee profile on inside edge

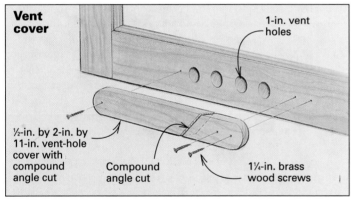

Vent cover

1-in. vent holes

½-in. by 2-in. by 11-in. vent-hole cover with compound angle cut

Compound angle cut

1¼-in. brass wood screws

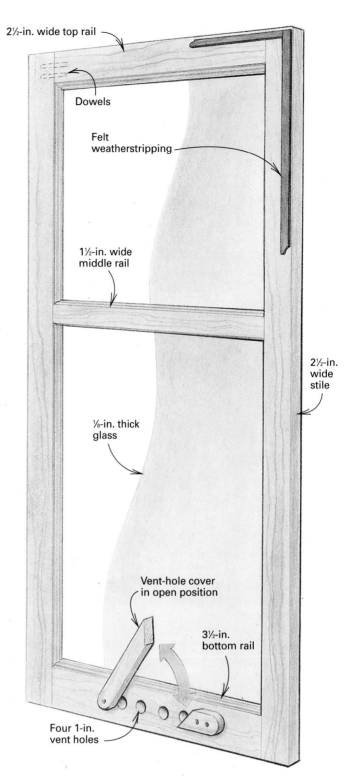

2½-in. wide top rail

Dowels

Felt weatherstripping

1½-in. wide middle rail

⅛-in. thick glass

2½-in. wide stile

Vent-hole cover in open position

3½-in. bottom rail

Four 1-in. vent holes

lot of stress, they would be moved around. That movement, combined with the weight of the two large panes of ⅛-in. glass they would bear, made doweled joints my choice.

Finally, I chose old-fashioned felt weatherstripping over foam-rubber weatherstripping for the insides of the storm windows. The felt wicks water away from the window, rather than absorbing it the way foam rubber does. Also, felt doesn't get brittle or flake off and so lasts longer than synthetics. The felt is available at most hardware stores.

The first step was measuring the opening and checking for square— Before I took out my tape to measure the window openings, I checked the top of the opening with a framing square. If the opening was out of square, I

marked the widest point and then measured from the sill to that mark to get the length measurement. I measured the width at the top, middle and bottom of the opening.

I used the widest and longest measurements to determine the size of the storm windows. I would do the final trimming once the windows were complete. After measuring all the windows and noting their peculiarities on a map of the house, I built a prototype window to help organize the sequence for building.

Each storm window would be divided by a rail at the same point on the window where the double-hung sash overlapped. Each window would have an ogee profile on the inside. Also, each would have four ventilation holes in the bottom rail, which would be covered with a closable wood hatch on the in-

side of the window. It's important to be able to vent storm windows to let moisture escape.

Starting a production run for 58 storm windows—Once I got the 350 bd. ft. of yellow poplar I needed, I rough-sized all the stock. The poplar was delivered in 12-ft. lengths, which enabled me to get two 6-ft. pieces out of each long plank. I crosscut all the longer lengths for the windows first; the heights of the two window sizes I would build were 70⅝ in. and 56½ in. The width of the outside rails and stiles was 2½ in. to match the window sash the storm windows would cover (drawing right, p. 107). Also, the frame needed that much width to support the weight of the glass.

Next, I went down my cutting list to determine the most efficient crosscuts for the remaining planks, which included cutting out 2½-in., 1½-in. and 3½-in. rail pieces for the top, center and bottom rails respectively.

I always end up with some warped stock, so I saved the most severely bowed pieces for the smaller rails. I set these pieces aside so that I could later joint out the warp. I would use the thinnest material for the ½-in. by 2-in. by 11-in. ventilation-hole covers (drawing bottom left, p. 107).

Finally, I joined one edge of each piece and ripped all of them to size. I made sure to keep all the stock crown side up and the stock organized by size while the windows awaited assembly.

Dowel holes are drilled before rails are coped—I use a Dowl-It jig (800-451-6872) to drill dowel holes. Using a 5⁄16-in. brad-point bit and a homemade stop collar, I can move right along with the joinery (top photo, p. 106).

At the same time I was marking out for the doweling, I marked out for the vent holes. I made up a dummy rail board that has a center mark and four other marks spaced 2 in. apart. From a center mark on the drill press, I could line up each of the four vent-hole marks and drill using a 1-in. Forstner bit.

After drilling all the dowel holes, I coped the ends of all the rail stock first. I used a back-up block against my miter gauge as I pushed each piece through the shaper (bottom photo, p. 106), which prevented the rails from splintering as they exited the cutter. After the rail stock was cut, I prepared the shaper for cutting the stile and rail edges. This entails lifting the bits off the spindle, installing a bushing that comes with the set, and reinstalling the ogee bit and rub collar. With this profile cut on the end of a rail, it meets the reversed profile on the stile, and the match is tight.

Waterproof polyurethane glue holds it together—Everything was ready for assembly. To glue the joints, I spread liquid polyurethane glue over them and swabbed glue into the dowel holes. I used polyurethane because it is waterproof, expands into crevices as it cures and retains some elasticity.

Polyurethane glue reacts to the moisture in wood. If the moisture content of the wood was below 8%, I dipped the dowels in water before inserting them in their holes. After inserting the dowels, I wet the edges to be glued with a damp rag, applied the glue and assembled the frame (photo above). Next I checked for square before pipe-clamping the window together.

Once the glue cured (two hours to five hours), I block-planed the small variations between the rails and stiles and cleaned up the ends. I cleaned both sides using a random-orbit sander and 120-grit paper.

Because of the ⅞-in. thickness of the windows, the rabbet from the sash cutter didn't leave as deep of a reveal as I needed for the ⅛-in. thick glass and the glazing compound. So I deepened the rabbet by using the rabbet bit in a plunge router fitted with a rub collar, cleaning corners with a butt chisel.

Making the vent-hole covers—Next I cut the stock for the vent-hole covers into ½-in. by 2-in. by 11-in. pieces. With a scratch awl, I marked and drilled the holes for the brass screws that hold the covers in place. I rough-cut the corners off each end of the vent-hole covers on the bandsaw. Then, using a belt sander in a holding jig and 120-grit paper, I finished rounding off the pieces.

Each vent-hole cover is two pieces divided by a compound angle that allows the two to mate together (photo top right). To make the compound cut on the vent-hole cover, I set my miter gauge at 52½° and the table-saw blade at 45°. I cut this compound angle 8⅜ in. from the end of the cover.

Placing both cover pieces over the vent holes equal distances from the ends and edges, I screwed the vent-hole cover into place with #6, 1¼-in. solid-brass wood screws. I took them off for painting and screwed them back in place after the windows were installed.

Prefitting comes before priming and glazing—It's a lot easier to prefit these units before the glass and glazing are in and paint applied. I started by getting my sawhorses, extension cord, circular saw with a sharp blade, power plane, block plane and ladder set up at the house.

I inserted the top of the window into the opening first (photo left). That way, the bottom of the storm window sticks out so that it can be marked for trimming off the 3½-in. wide bottom piece. A bit sometimes had to come off each side, depending on how close I had made them.

To get the bevel I needed on the bottom of the storm window, I set my bevel square against the blind stop at the windowsill. Then I transferred this angle to the bottom edge of the storm window. I adjusted my circular saw to match the angle on the windowsill and made my cut.

After prefitting the window, I took it out and primed it. Depending on the type of glazing compound that I used, I applied either linseed oil or latex primer (for more on glazing windows, see *FHB* #99, pp. 65-67). Because I had decided to use a polyurethane glazing caulk, I primed the windows with a latex primer.

At this point I took the windows to my local glazier. After cutting the glass, he ran a flat bead of polyurethane glazing caulk onto the rabbet and pressed the glass into this before pointing in the glass. Polyurethane glazing caulk adheres to any dry surface and dries to a pliable rubber consistency that won't crack. Next, the glazier used the caulk gun to run a flat bead around the window where the glass meets the frame (photo bottom right).

After applying a top coat of paint, I nailed felt weatherstripping around the outer edge on the inside of the storm windows. The windows were then positioned into their openings. I used four zinc-plated steel turn buttons to hold each window in place. □

Rex Alexander, a custom woodworker who lives in Brethren, Michigan, specializes in cabinets and staircases. Photos by Steve Culpepper.

Fitting the windows. Positioning the top of the window first shows how much has to be trimmed from the sides and the bottom rail, which is 3½ in. deep to accommodate the windowsill bevel.

A compound angle ensures a tight fit. Shaped from an 11-in. piece of poplar and cut at a compound angle, the small end of the vent-hole cover is secured with two screws to the inside of the storm window.

Polyurethane glazing caulk means fast glazing. After the glazier pointed the glass in place, he glazed the window in minutes using a caulk gun and a tube of polyurethane window-glass caulk.

Sizing Up Housewraps

Installed correctly beneath exterior siding, these tough plastic fabrics can help cut
energy bills and provide a hedge against the weather

Better than 15-lb. felt? Although felt is less expensive, housewraps come in 9-ft.
wide rolls, which means fewer seams and faster installation.

by Bruce Greenlaw

"I don't use a building material unless it's 100 years old," says Art Ward, a contractor in the San Francisco Bay area. Ward's company, Ward Construction Inc. specializes in repairing structural damage to homes. Over the years, Ward has encountered countless problems resulting from the use (or misuse) of building materials that haven't passed the test of time. When conditions call for the installation of a weather-resistant building paper on exterior walls, Ward uses asphalt-saturated felt, not housewrap.

Housewraps are thin, lightweight plastic fabrics that measure as large as 10 ft. wide and come on a roll, like wrapping paper. They're stapled or nailed to sheathing or to bare studs, just ahead of the siding (photo above). Seams and edges are sometimes taped or caulked. Installed properly, housewraps are supposed to help trim energy bills and prevent moisture problems.

The 10 housewraps now on the market don't even come close to meeting Ward's criterion. Tyvek, the first housewrap, was introduced by Du Pont in the late 1970's. Building pros still debate the merits of housewrap. Some even argue that well-built houses don't need housewrap.

Is it worth spending about 6¢ to 8¢ per sq. ft. for a housewrap, plus installation costs? If so, how do you install it to maximum advantage? To find out, I pestered manufacturers, builders, architects, building scientists, trade associations

and building-code agencies throughout the United States and Canada. I learned that you can't staple up a housewrap and expect it to work miracles. But housewraps do appear to work well in many exterior-wall systems, especially if you heed details such as window flashing and seam sealing. Incidentally, I learned that asphalt-saturated felts were once made of recycled paper, sawdust and cotton rags. But according to Tamko Asphalt Products, a felt manufacturer, cotton is no longer used. So even old building materials can change when nobody's looking.

High-tech plastics—Housewraps are often compared to Gore-Tex. They're designed to repel rain, to keep out the wind and to breathe (or allow water vapor to pass through them), just what you'd expect from a good Gore-Tex raincoat. All housewraps on the market are made of either polyethylene or polypropylene (or polyolefin, a generic term for the same class of materials). Unlike garden-variety poly sheeting, which is a homogeneous extruded film, housewraps are spun, woven, laminated or fiber-reinforced for tensile strength and resistance to tearing. Some housewraps are coated with polyethylene or polypropylene for structural reinforcement and resistance to air leakage. Others don't need it. Two products on the market, Tyvek and R-Wrap, transmit water vapor between their fibers while repelling bulk water. The rest don't, so they're perforated with tiny holes that are engineered to let water vapor pass through while keeping liquid water out. (For more comparisons, see "Choosing Housewraps," p. 114.)

Housewraps offer some distinct advantages over asphalt-saturated felt. For example, R-Wrap, the heaviest housewrap, weighs just 17.3 lb. per 1,000 sq. ft., while Tamko 15-lb. asphalt-saturated felt weighs 135 lb. per 1,000 sq. ft. A housewrap's light weight makes little difference to a building, but it can make life easier for a carpenter. Most housewraps also are more rugged than 15-lb. felt, so they're less prone to damage on job sites. And, unlike felt, housewraps come in wide rolls that can cover a one-story house in a single pass for speedy installation. Housewraps remain flexible in subfreezing temperatures, too, allowing cold-weather handling. Perhaps most important, housewraps make better air retarders than felt because their seams and edges can be sealed easily. Felt, on the other hand, is cheaper (about 2¢ to 4¢ per sq. ft.), is available everywhere and has an excellent track record as a rain barrier.

Housewraps and building codes—Model building codes in the United States generally require the installation of 15-lb. asphalt-saturated felt or an equivalent weather-resistant building paper on wood-frame exterior walls behind shakes, shingles, board sidings, horizontally applied siding panels, stucco and masonry veneers such as brick and stone. Felt or its equivalent usually isn't required under vertically applied panel sidings. And, surprisingly, felt or its equivalent usually isn't required over water-repellent panel sheathings such as CDX plywood, OSB (oriented strand board), exterior gypsum sheathing or extruded-polystyrene foam insulation.

Trying to figure out which housewraps can be substitutes for 15-lb. felt can be a baffling subject for builders. It also can be puzzling for many building inspectors.

Every type of housewrap is listed by a code organization somewhere as an alternative to 15-lb. felt for resisting the penetration of water. This means that manufacturers have paid a hefty fee to a particular code agency and have supplied the agency with pertinent data from tests by independent labs. A building inspector can okay the use of an unlisted housewrap, but the burden is on the builder to convince the inspector that it's an effective substitute either for felt or for a listed housewrap.

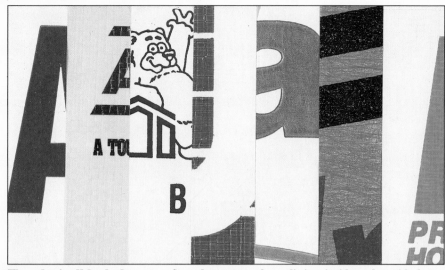

They don't all look the same. Some housewraps have distinct inside and outside faces; others don't. Semitransparent housewraps let you see the substrate through them, which helps when installing siding. Darker housewraps are easier on the eyes in bright sunlight.

Housewrap-installation pointers

- **All housewraps aren't alike.** For instance, some housewraps have a distinct inside and outside face, and others don't. Also, some require the use of more or larger fasteners, or both, than others do. Be sure to read the manufacturers' instructions for installing housewraps.

- **Housewraps are slippery.** If you have to lean a ladder against one, mount rubber ladder caps or a top-mounted, rubber-tipped standoff stabilizer on the ladder to help in prevention of sliding.

- **Siding contractors often work off of pump jacks or scaffolding.** If yours do, consider having them tape the housewrap seams and perhaps even install the housewrap.

- **Regardless of their exposure ratings,** all housewraps start to degrade somewhat as soon as they see sunlight.

Even Typar, which boasts unlimited exposure time, can lose about 20% of its tear strength after 16 weeks of UV exposure (according to Reemay's own literature). For this reason, it's a good idea to cover up housewraps with siding as soon as possible.

- **Use sheathing tapes, not duct tapes** or other packing tapes, to seal the seams of housewraps. Sheathing tapes are expensive, but they're designed to stick to the housewraps for the life of the building.

- **Construction adhesives, latex sealants, polyurethane sealants and even acoustical sealant are all used as housewrap adhesives.** Before applying a sealant or an adhesive to a housewrap, though, check with manufacturers for compatibility.
  —*B. G.*

Conflicting opinions about housewrap as a substitute for felt—I called several siding associations and manufacturers, all of which gave housewraps "thumbs up" as substitutes for 15-lb. felt. Jeff Fantozzi, special-products manager of the Western Wood Products Association, says his association recommends the use of a housewrap or other sheathing paper under all wood-board sidings regardless of sheathing used or building-code requirements. Ed Keith of the American Plywood Association said he considers housewrap a cheap insurance against the weather, even where it isn't required by code. And Alcoa, which makes vinyl, aluminum and steel siding, sells Tyvek housewrap as an accessory, even

though most codes don't require a sheathing paper under vinyl, steel or aluminum siding.

Despite the preponderance of recommendations in favor of housewraps, John Ward (Art Ward's father) is still unconvinced. Ward is vice president of Ward Construction Inc. and a construction consultant for M.J. Ward and Associates, which specializes in diagnosing construction problems such as water leaks. He tells me that asphalt-saturated felt keeps out errant water better than housewraps because felt seals itself to staples or nails that penetrate it, while at least some housewraps don't.

To find water leaks in houses, Ward and his associates typically spray water with a garden hose

Tape the plywood, skip the housewrap. Gap Wrap Air Infiltration Barrier is designed to replace housewrap. It's an adhesive-backed polypropylene tape that covers sheathing seams and other exterior gaps to help prevent air infiltration. Unlike most housewraps, though, it doesn't replace 15-lb. felt where the use of a water-resistant sheathing paper is required by either the building code or by common sense. Photo courtesy Benjamin Obdyke Inc.

Sealed seams, better results. For maximum resistance to air infiltration, gaps must taped or caulked. Here, a 3M sealing-tape dispenser is used on seams around windows. Photo courtesy of 3M Construction Markets Division.

on suspect areas, working their way from the bottom of a wall up onto the roof if necessary until they find the problem. Several times, they've discovered water leakage through housewraps (specifically Tyvek and Barricade) where fasteners pass through it. Joe Lstiburek, principal of Building Science Corporation in Chestnut Hill, Mass., and an expert on moisture and energy issues, tells me that he has seen the same thing with some housewraps. Ward says that although asphalt-saturated felt tends to get brittle with age inside walls, he has never seen it leak this way when subjected to the same test. He replaced the housewrap on the leaky walls with felt.

Keep in mind that the leaky houses that Ward mentions are located on hillsides pummeled for days with windblown rain during winter storms. Also, neither housewraps nor felts are designed to be primary rain barriers. If an appropriate siding is chosen for the job, if the siding is installed right, and if windows and doors are flashed carefully, windblown water shouldn't be a problem.

A wet-climate strategy—Fortunately, installation techniques have been devised that help keep housewraps and substrates dry, even in extreme climates. The Canadian Wood Council recommends the construction of "rain-screen" walls (left drawing, facing page). Similar in concept to the typical brick-veneered, wood-frame wall, a rainscreen is simply a wall with a vented airspace behind the siding. The airspace is formed by nailing 1x furring strips either horizontally (with drain holes drilled through them) or vertically over the housewrap, depending on the orientation of the siding. The airspace is typically blocked off at the top and the sides, but it is left open at the bottom. Insect screen keeps bugs

out. The airspace not only provides a capillary break that prevents water from wicking through, but it also equalizes air pressure on both sides of the siding during windy, rainy weather, preventing a pressure differential that can drive water through even small holes in siding. Rainscreen walls also help dry wet siding quickly and evenly, preventing cupping and other problems.

Flashing housewraps around windows—Most housewrap manufacturers recommend that you install the housewrap, cut an "X" in the rough window openings, fold the flaps into the openings and nail or staple them flat. Then you install the windows. But windows must be flashed right, or you might as well throw the rest of your waterproofing strategies out of one.

The easiest flashing solution (right photo above) is simply to seal the window's nailing flanges to the housewrap with an approved tape such 3M 8086 Builders' Sealing Tape (3M Construction Markets Division, 3M Center, Building 225-4S-08, St. Paul, Minn. 55144-1000; 612-736-2388). This requires that you install the tape without wrinkles and then depend on its adhesive to keep out water for the life of the building. Or you can squash some caulk behind the window flanges instead. Both methods break the rule that you should always use geometry to shed water. Nevertheless, they probably work fine for windows that get little or no exposure to rain.

James Larson, an architect and consultant in St. Paul, Minn., has seen some real disasters due to botched window flashing. He suggests adding a couple of extra steps (middle drawing, facing page). First, he recommends slitting housewrap horizontally just above rough window openings to allow head flanges to slip underneath while

the rest of the flanges sit on top. The flanges are then taped to the housewrap. The final step is to inject a small bead of foam sealant between rough openings and window frames from inside the house using a can of foam with a long spout. This sealant provides a backup air retarder and water repellent.

Air control—Despite the importance of repelling bulk water, the biggest selling point of housewraps is their ability to retard air infiltration. Air leaks make it tough to heat and cool a house; air also can transport an amazing amount of moisture into exterior walls. According to the Canadian Wood Council, air propelled by pressure equal to a 10-mph wind can haul 30 lb. of water vapor per month into a wall through a 1 sq. in. hole. If moist air is cooled below its dew point in there, the resulting condensation can be the cause of mold, mildew or rot. Generally, air exfiltration creates moisture problems in heating climates, and air infiltration causes them in humid, cooling climates (for a more detailed explanation, see "Air and Vapor Barriers" in FHB #88, pp. 48-53).

A current study by Du Pont on several two-story colonials in Charlottesville, Va., has found that simply stapling Tyvek housewrap to the wall sheathing has reduced air leakage by about 14%. Adding Headerwrap (18-in. wide strips of Tyvek that thread under mudsills and between the top-floor top plates) and taping all the seams has cut air leakage by almost 30%. Keep in mind that housewraps seal walls only, so they're just one piece of the puzzle.

Housewraps help maintain insulation's R-value—There's at least one more good reason

In wet climates.
Rain-screen walls feature a vented airspace that balances air pressure on both sides of the siding during stormy weather to repel rain and help keep housewraps drier than conventional walls do.

Stud

Sheathing

Housewrap

³⁄₈-in. to ³⁄₄-in. thick furring strips

Airspace vented at bottom of wall

Siding

Insect screen installed over bottom of airspace

Around windows.
To prevent water leakage around windows, James Larson, an architect in St. Paul, Minn., suggests tucking head flanges under housewrap and taping the perimeters. Instead of taping, some builders squash a bead of compatible caulk between housewrap and flanges.

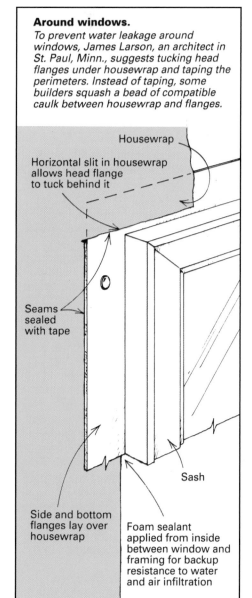

Housewrap

Horizontal slit in housewrap allows head flange to tuck behind it

Seams sealed with tape

Side and bottom flanges lay over housewrap

Sash

Foam sealant applied from inside between window and framing for backup resistance to water and air infiltration

At seams and edges.
The top-of-the-line housewrap job calls for the installation of narrow strips of housewrap between the mudsill and a sill sealer, as well as between the top-floor top plates. Walls are wrapped with upper courses of housewrap overlapping lower ones for drainage. Seams, tears and other gaps are sealed with an approved tape. As an alternative, the bottom edge of the housewrap can be caulked to the foundation or taped to the mudsill. The top edge is taped to the top-floor top plate.

Strip of housewrap installed between top plates

Housewrap

Tape all seams.

Strip of housewrap installed between sill sealer and mudsill

to install housewraps. Research has shown that wind can take a hefty bite out of an insulation's R-value, a phenomenon called "wind washing." But a recent study done for Du Pont by Holometrix Inc., an independent laboratory in Bedford, Mass., found that housewraps can help prevent wind washing even on walls that are tightly built.

Holometrix framed a wall with 2x4s, insulated it with R-13 kraft-faced batts and applied ½-in. drywall to one side (sealed using the airtight drywall approach) and ½-in. extruded-polystyrene foam sheathing to the other. The wall then was exposed to a simulated 10-mph wind. The wall's thermal performance plummeted 19%. With Tyvek installed (complete with Headerwrap and taped seams), thermal performance dropped just 1%. Depending on the brand of housewrap used, housewraps improved thermal performance by 57% to 247% when test walls were exposed to simulated winds ranging from about 9 mph to 14½ mph. Tyvek worked best. (At this point, most of

the research that I know of on housewraps has been funded by Du Pont.)

Air-proofing strategies—If you want your housewrap to act as a secondary rain barrier and a superficial air retarder, you can staple or nail it up as most builders do (for basic installation tips, see sidebar, p. 111). This requires that all vertical seams be lapped at least a few inches and that all horizontal seams be lapped shingle-style (the upper course overlaps the lower one). Mudsills should be lapped, too, and the housewrap should extend up to the top-floor top plate. Gaps around windows and doors should be caulked or taped.

If you want a housewrap to work as an effective air retarder, though, you'll need to tape all the seams and holes in it, too. To do this, you need a sheathing tape designed especially for this purpose. You'll also need to tape or caulk the top edge of the housewrap to the top-floor top plate and the bottom edge to the mudsill. Or, better yet, you can lap the mudsill and caulk the housewrap to the foundation.

An alternate method (right drawing, p. 113) requires installation in two stages but eliminates taping or caulking to wood or concrete, which can produce an unreliable seal. First, a narrow strip of housewrap (such as Du Pont's 18-in. wide Headerwrap) is installed between the mudsill and a sill sealer. Later, this strip is bent upward and fastened to the sheathing (or to the studs if there is no sheathing). A second strip of housewrap is installed between the top-floor top plates (not over them, because housewraps are too slippery to stand on while roof-framing). The rest of the house is then wrapped shingle-style so that upper courses overlap lower ones for drainage. Finally, all joints are taped with sheathing tape. No doubt, installing these extra strips of housewrap is initially more hassle for builders because it's done during the framing stage, but it does make sealing seams easier.

One manufacturer, Benjamin Obdyke Inc. (John Fitch Industrial Park, Warminster, Pa. 18974; 800-346-7655), suggests that you can skip housewrap entirely and control air leakage by simply taping the seams and edges of the sheathing. Its product, Gap Wrap Air Infiltration Barrier (left photo, p. 112), is designed for this purpose.

Would I put housewrap on my house?—It depends. In the dry desert climate where I live, the answer is no. Housewraps have limited value in warm, dry climates, where moisture vapor problems are almost nonexistent. As for controlling wind washing, most areas of my county don't even require insulation.

If I lived in a hot, humid climate such as Florida's, I might heed the advice of the Florida Solar Energy Center and install an exterior air retarder such as a housewrap. This would help prevent outdoor humidity from condensing on air-conditioned surfaces inside the building envelope, but only if the seams and the edges were sealed.

In a heating climate, I'd consider installing a fully sealed housewrap (including Headerwrap) to serve as the primary air barrier. Or I might use housewrap for sealing complicated exterior details such as the bottoms of cantilevered rooms and bays, taping or caulking the edges of the housewrap to an interior air-vapor retarder. Mark LaLiberte, an energy consultant and co-owner of Minneapolis-based Shelter Supply, says this works great.

If I need a secondary rain barrier without concern for air infiltration, I'll use 15-lb. felt. As for slapping a layer of housewrap on my house for peace of mind as many builders do, I'd rather use the money to buy a new biscuit joiner and to go out to dinner with the change. □

Bruce Greenlaw is a contributing editor of Fine Homebuilding. He divides his time between Bisbee, Ariz., and the High Sierra of central California.

Choosing Housewraps

Unlike cars, refrigerators and air conditioners, housewraps aren't held to a uniform standard, so comparing brands isn't easy

Want to compare the performances of two air conditioners? Check their SEER ratings (seasonal energy-efficiency ratios). Want to compare the performances of two housewraps? Good luck. Housewrap manufacturers provide specifications all right, including those for breaking and tearing strength, water-penetration resistance, air-penetration resistance and water-vapor transmission—everything you need to know. Trouble is, manufacturers don't use all the same performance tests, so it's hard to compare housewraps head to head for each specification. For example, Du Pont claims that Tyvek resists air infiltration better than ProWrap, based on a test called ASTM E-283. ABTCO uses ASTM test D-726 to show that ProWrap resists air infiltration better than Tyvek. So what's a builder to do? Fortunately, help might be on the way.

A standard rating system—Mike McKitrick, chairman of an ASTM (American Society for Testing and Materials) committee that's charged with creating a new standard for specifications of air retarders, tells me a new standard could be forthcoming soon. If adopted, this standard will address all air retarders for walls, including housewraps.

To comply to the ASTM standard (compliance is voluntary), air retarders will have to meet minimum performance requirements for air leakage, structural integrity, bulk-water resistance and water-vapor resistance.

For builders, the beauty of the ASTM standard is that air retarders will be tested in a wall according to the manufacturer's published installation instructions, not as stand-alone items. Builders will get a better idea of relative performance, and they'll also be able to shop for the housewrap that works best for a particular application, such as for repelling wind-blown rain.

Until the ASTM standard is finalized and subsequently adopted by manufacturers, comparing housewraps will continue to be difficult. But after poring over mounds of product literature and manhandling samples of all the products, I will take a venture with some observations.

Housewraps will hold water—Based on a standard ASTM test called E-96, Procedure A, R-Wrap (the new kid on the block) claims to have the highest water-vapor-transmission rate. This means that it's the least likely to trap water vapor in wall cavities and cause condensation problems.

Tyvek is rated a distant second, but it is well above the rest of the pack (including 15-lb. felt).

Du Pont uses the "hydrostatic head" test to compare the relative resistance of

housewraps to liquid-water penetration. Samples are placed horizontally under an open cylinder, which is then filled with water to a height that causes the sample to leak. The higher the column of water, the better a sample is supposed to resist liquid water such as wind-blown rain. No housewrap is waterproof, but Simplex claims that R-Wrap holds more than 186 cm of water before it leaks. According to Du Pont, Tyvek holds more than 130 cm. Airtight-Wrap and Tu-Tuf Air Seal hold 11 cm. Based on independent-lab tests, Du Pont claims that the rest hold between 11 cm and 15.3 cm. (For comparison, Tamko 15-lb. felt holds more than 50 cm.) It can be argued that this test is worthless because it doesn't consider the effects of fasteners and those other real-world factors. But it does give a general idea of the relative water-resistance of the fabrics themselves.

Truth is, it's hard to believe that the perforated housewraps will hold any water. So I poured a pool of water from my kitchen faucet into each of my housewrap samples. Sure enough, none of them leaked.

Air resistance and strength—Specifications for resistance to air leakage are smoggy at best. Based on tests a few years ago by an independent Canadian lab called Air-Ins Inc., Tyvek is slightly more resistant to air infiltration than 15-lb. felt and significantly more resistant than Barricade, Typar and those housewraps made of Valeron (ProWrap, Airtight-Wrap, Rufco-Wrap, Tu-Tuf Air Seal), a chemical packaging material manufactured by Van Leer Flexibles Inc. Simplex claims that its product, R-Wrap, is better than Tyvek, but Simplex hasn't used the same test to prove it. (Keep in mind that these are tests of the materials themselves and not of the materials under installed conditions, where the seams and the edges of the housewraps are much easier to seal than felt.)

Strength specifications are not just inconsistent, they're deceiving. For instance, specs might show that a housewrap won't break or tear easily. But it might stretch and rupture, which is just as bad. To get some idea of relative strength, I grabbed opposite ends of each of my housewrap samples and yanked hard several times. I did this lengthwise, widthwise and diagonally. Tuff Wrap and Barricade, both woven polyethylenes, popped open all over the place when I pulled diagonally, creating gaps big enough to peek through. None of the other housewraps did this.

On the other hand, woven housewraps appear to be highly tear-resistant. I made knife cuts in two perpendicular edges of each sample, then tried to propagate a tear

Housewrap Specifications									
Manufacturer	Brand name	Material	Stock sizes (in ft.)	Thick-ness (mils)	Weight (lb./1,000 sq. ft.)	Color	Semi-trans-parent	Max. exposure time*	Notes
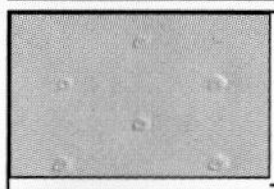 ABTCO, Inc 800-334-3551	ProWrap	Perforated, cross-laminated poly-ethylene (Valeron)	9×111 9×200	3	14	white	yes	6	Sells ProWrap sealing tape
 Amoco Foam Products Co. 800-241-4402	AMO-WRAP	Perforated, polyole-fin-coated, woven polypropylene	4½×195 9×100 9×195	8	17.6	green	yes	12	
 The Celotex Corp. 813-873-1700	Tuff Wrap	Perforated, poly-ethylene-coated, woven polyethylene	4½×195 9×195	5	17	white	yes	3	
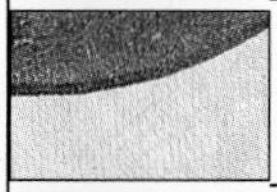 Du Pont 800-448-9835	Tyvek	Spun-bonded polyethylene	1½×100, 3×100, 3×165, 5×200, 9×100, 9×195	3.77--6.44	10.6	white	no	4	Has stud marks spaced 8-in. o.c.
 Parsec, Inc. 800-527-3454	Airtight-Wrap	Perforated, cross-laminated poly-ethylene (Valeron)	4½×300, 9×60 9×195, 9×107	3.7	12.7	blue	yes	4	Custom sizes available. Sells vapor and radiant barriers, sealing tape.
 Raven Industries, Inc. 800-635-3456	Rufco-Wrap	Perforated, cross-laminated poly-ethylene (Valeron)	4½×111 4½×195 9x111, 9x195	3	14	white	yes	6	Sells TS-50 Transit Seal tape.
 Reemay, Inc. 800-321-6271	Typar	Perforated, spun-bonded, polypropylene-coated polypropylene	3×111 4½×222 9×111	11-5	16.5	platinum	no	unlimited	Has stud marks spaced 8-in. o.c.
 Simplex Products Div. 517-263-8881	Barri-cade	Perforated, polyethy-lene-coated, woven polyethylene	3×195, 4½×100 4½×195, 9×110, 9×195	5	14	white	yes	12	
 Simplex Products Div. 517-263-8881	R-Wrap	Non-perforated, fiber-reinforced polyolefin	3×150, 4½×100 4½×150, 9×100 9×150	9	17.3	white	yes	4	
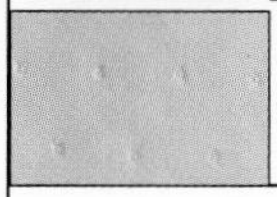 Sto-Cote Products, Inc. 800-435-2621	Tu-Tuf Air Seal	Perforated, cross-laminated poly-ethylene (Valeron)	4½×100, 4½×195, 9×100 9×195	3.4	15	white	yes	12	Lay-flat or J-fold rolls. Custom widths and sealing tape available.

The number of months that you can leave a housewrap exposed on a building without serious degradation due to ultraviolet rays.

Of the 10 brands on the market, some are more water-resistant, others hold up better in the sunlight, and one can be printed with your company logo

at each cut by pulling on the fabric. AMOWRAP was the toughest here. Tuff Wrap, Barricade and R-Wrap all resisted tearing, too (the woven strands on the Barricade and the Tuff Wrap did begin to pull apart, though). Valeron housewraps all tore fairly easily, with top and bottom layers tearing diagonally in opposite directions and delaminating (9-ft. wide ProWrap—a Valeron housewrap— has a "staple strip" in the middle for added strength and resistance to tearing). Tyvek and Typar tore fairly easily, too. This might be why Du Pont recommends the use of big (1-in. crown) staples or nails with big heads or plastic washers on them.

Colors, UV resistance and housewraps with your company logo on them—
Beyond performance characteristics, housewraps have other qualities worth comparing. Semitransparent housewraps, for instance, are the easiest to install because you can see substrate voids through them, reducing the chance that you'll puncture them accidentally with a hammer or a staple gun.

They make siding installation easier because you can see stud locations through them.

The color of a housewrap can be important in some climates. AMOWRAP is green, and Typar is a platinum color. Both are easier on the eyes in bright sunlight than white housewraps, which tend to glare.

And, speaking of sunlight, resistance to ultraviolet radiation can vary for different housewraps. Tuff Wrap, for example, is supposed to be covered by siding within three months, while Typar supposedly can be left uncovered indefinitely. If your housewrap will be exposed for awhile, you should probably take UV resistance into consideration.

If you plan to install a housewrap in the Alaska Range in mid-January, a Valeron one

might be a good choice. Valeron remains pliable down to at least -70°F.

Most housewraps come unfolded in rolls, but Tu-Tuf Air Seal and Rufco-Wrap are also available in J-fold rolls, which provide 9-ft. wide sheeting on a 5-ft. wide roll. Their manufacturers claim that J-fold rolls are the easiest to install on vertical walls.

Some housewraps come in more sizes than others do, giving more installation options. Du Pont offers the best selection of stock sizes, but Parsec sells Airtight-Wrap in custom sizes as large as 16-ft. wide by 1,000-ft. long and Sto-Cote sells Tu-Tuf Air Seal in custom widths in 4½-ft. increments. Some companies also sell special tapes for sealing seams and patching holes in their housewrap. Others don't. Sealing tapes are readily available for the other housewraps, though.

If you order at least 32 rolls of Barricade, you can get your company name and logo printed on it for about $3 per roll. Finally, the cost of housewraps can vary as much as about 2¢ per sq. ft. Call around for prices in your area.—*B. G.*

Installing Housewrap

When properly detailed, high-tech wrappings can reduce the flow of air into outside walls while allowing moisture to escape

by Rick Arnold and Mike Guertin

Tape completes the air barrier. Housewrap just stapled to the exterior sheathing does not create a complete air barrier. All seams and holes must be sealed with specialized tape before the siding goes on.

Build tight, ventilate right. A slogan with roots in energy-efficient building is gaining acceptance in mainstream construction. More builders are installing better insulation and sealing the warm side of walls and floors to keep vapor out of insulated spaces. They're installing ventilation systems to maintain healthful indoor-air quality. But from what we've seen, the same care isn't taken with air-infiltration retarders, also known as air barriers or housewraps.

There are usually a couple of ways to do something right and dozens of ways to do it wrong. Most of the housewrap installations we've seen in our area fall into the latter category. Sure, the walls might seem covered, but there are enough seams, gaps, tears and holes at critical spots to render the film ineffective (photo above). We asked some builders why they use air-infiltration retarders, and we got some interesting answers: "What's an air-infiltration retarder?"

"That's what's spec-ed on the plans." "It's better than tar paper." "I only use it because customers expect it." "It makes the house waterproof." "It dries in the building until the siding goes on." Because we know what housewraps are supposed to do, we apply the details that will make the barriers most effective. We've also learned how to coordinate installation with our framing habits to make installation easy. We've installed every major brand of housewrap and have

Compressed air is quicker than a broom. Where possible, the authors install housewrap on walls before the walls are raised. Compressed air is used to blow wood chips and other debris off a wall before the wrap is rolled out.

A two-man team makes the work go faster. Especially on windy days, it's good to have one person stapling housewrap while another unrolls the material.

Overlapping sheets create a good seal. When installing housewrap on a new wall, the authors fold up the bottom foot or so of housewrap to overlap the wrapped wall below.

found little or no difference in installation. Although Tyvek was specified for this project, we usually request Typar because its gray color is easier on the eyes than bright white housewraps.

Housewraps are an effective product when installed correctly—Air-infiltration retarders enhance the thermal efficiency of exterior walls by reducing air movement through walls and into wall cavities. A properly installed wrap slows or stops wind- or pressure-driven air from moving freely through gaps and holes in sheathing or around window frames and door jambs. Reducing drafts makes a house feel more comfortable and saves energy. At the same time, housewraps allow moisture vapor that enters the wall cavity to escape. Without this feature, moisture could build up, and rot could begin inside the wall.

Before housewraps entered the market, we used 15-lb. tar paper or red rosin paper beneath our siding. We surmise now that tar paper did a pretty good job of reducing air infiltration and was waterproof.

On occasion, tar paper caused us problems. Stains bled through wood siding, and dissolved tar leaked onto vinyl siding. During remodeling jobs, we sometimes find concealed rot caused by years of condensation behind tar paper.

Rosin paper probably helped to reduce some air infiltration and to let moisture escape. But it disintegrated if it got wet before siding was installed or if small leaks occurred around windows and along cornerboards. On remodeling

Tying up the loose end. At one end of the wall, the wrap is folded over the corner stud and stapled. After this wall is raised, sheathing from the adjoining wall will be nailed over the corner, holding the wrap in place.

Corners are the last to be wrapped. On the opposite end of the wall, housewrap that earlier had been rolled up on a furring strip is unrolled and fastened in place, which happens after the framing is tied together with the final pieces of sheathing.

jobs we have uncovered old rosin paper that had practically turned to dust.

Follow the directions and use common sense—The instructions supplied by housewrap manufacturers range from cursory to highly detailed. The one thing most companies include is a warning about securing ladders leaning against wrapped walls. The top of an unsecured ladder can slide sideways across the slick material. Our approach to installing air-infiltration retarders combines common sense and those recommended installation details that we believe are feasible.

Our system begins with a foam sill seal between the foundation and the mudsill. If every effort is being made to create a tight seal, we caulk the bottom edge of the wall sheathing to the mudsill to prevent infiltration at that point.

We once installed a 2-ft. strip of housewrap between the sill-seal foam and the mudsill, according to the instructions of one manufacturer. The idea was to lift this flap and wrap it over the band joist after we raised the exterior walls. We found this detail frustrating; the material blew around on all but the calmest of days. And we cringed as the flap was shredded by cords, toolbelts and hoses before the walls were up. In special cases, such as when we're building a wall with siding applied directly to the studs, we still might use this detail. However, most manufacturers' instructions say to staple the housewrap to the bottom edge of the sheathing and to cut the housewrap flush with the bottom of the sheathing. The first course of siding should hold the housewrap tight against the sheathing.

Start with a clean wall—As a rule, we lay out, frame and sheathe exterior walls before lifting them. So it makes sense for us to wrap them as much as possible before they are lifted.

First we sweep or blow off debris on the wall (photo left, p. 117). Anything we miss will leave lumps for the siding installers. One crew member positions the roll of housewrap on the already-sheathed wall. A second person holds the starting edge and staples the housewrap to the sheathing lining up the housewrap just by eye (photo top right, p. 117). Because the sheathing extends beyond the bottom plate to cover the band joist, we don't staple the lowest 12 in. of housewrap (bottom photo, p. 117). This allows us to lift the wrap and nail off the sheathing later without punching hundreds of holes in the wrap. Stapling high also makes it easier to fold up any housewrap from the wall or sill beneath.

We roll out about 10 ft. before stapling off the sheet. It's important to tack the first corner and then to pull the wrap tight in all directions. Once the first 10 ft. are well-tacked, we roll out 3 ft. to 4 ft. at a time, stapling along the edge of the roll

Housewrap is cut to fit framed window openings. Instead of the standard X-cut for a window opening, the housewrap is cut to cover just the framing to eliminate large flaps that are likely to tear in the wind.

Sealing the corners. The authors try to tape the corner seams of the housewrap as soon as possible after the wrap is installed to prevent any possible wind damage.

Extending your reach. For sections of housewrap that are hard to reach, such as between-floor seams, a hammer-stapler is taped to a furring strip to make fastening much easier.

Tape discourages air infiltration around door and window openings. On windows and doors with nailing flanges, tape is applied so that it seals the seam between the flange and the housewrap.

of housewrap. Staples are spaced every 16 in. along the edges and every 24 in. in the field.

On one end of the wall, we leave the air barrier about 4 in. long and then wrap it and staple it to the corner stud (top photo, p. 119). Later, when the walls are raised, this 4-in. overlap will be covered and held in place by the sheathing from the other wall that makes up that corner.

When framing walls, we typically leave off the last 16 in. to 48 in. of sheathing on one end so that once the walls have been raised, we can come back and tie together the outside corners. On this end of the wall, we roll out enough

housewrap to cover these unsheathed sections and to extend 1 ft. around the corner of the house. The loose wrap is rolled around an 8 ft. furring stick, which is then tacked to the wall. This keeps the housewrap from blowing in the wind and being torn off. After lifting the walls, we nail on the final pieces of corner sheathing, unroll the housewrap and staple it around the corner (bottom photo, p. 119).

While the walls are on the deck, we take care not to step onto a housewrap-covered door or window opening. You can't fall any farther than the thickness of the wall, but a step on a win-

dow opening will rip out the surrounding staples in the housewrap.

When we have to install housewrap after the walls are lifted, it's much tougher to keep the job straight and wrinkle-free because someone has to be holding up the roll, maintaining tension on the wrap and unrolling the wrap all at the same time. Inside corners also can be a challenge: You must be careful not to short-sheet the corner or not leave enough material to go all the way into the corner. Short-sheeting can be avoided by pressing the housewrap into the corner with a furring strip before it is stapled.

More than one way to handle openings—At door and window openings, manufacturers suggest that an X-shape be cut in the housewrap with the slashes beginning in each corner of a door or window. The problem with this approach, we've found, is that you're left with large triangular pieces flapping in the breeze.

To prevent this, we feel for the edges of the openings, staple them off and then cut a rectangular piece out of the opening, about 3 in. smaller than the opening itself. We save the large rectangular pieces we cut out to fill in here and there. Staples fastened about every 16 in. to 24 in. at openings prevent the wind from tearing off the wrap (top photo, p. 119). The housewrap photographed for this article survived tropical storm Bertha unscathed.

Second-floor wrap overhangs the first—Second-floor walls get the same basic treatment as first-floor walls. We make sure to leave enough housewrap overhanging the bottom edge of the sheathing to lap the first-floor housewrap by 6 in. to 8 in. We usually have to fill in a foot or two of housewrap at the top of the wall with strips from the roll or with the pieces we saved from the window cutouts.

When you're installing housewrap, it doesn't matter whether the writing is right side up or upside down, but air barriers do have an inside and an outside. Manufacturers don't tell you to install their wraps with the printed side facing out just for free advertising. (We've been waiting for some town to impose a fine either on the housewrap manufacturer or the builder for violating a local sign ordinance. We can't put a small sign in the yard to advertise our company, but Tyvek can turn a whole house into a billboard.) The material's performance may be diminished if it's installed inside out.

We handle the intersection of walls and roofs in a couple of different ways, depending on the soffit detail. On a house with a trussed roof, we cut the housewrap along the top of the second-floor top plate and nail a soffit nailer to the wall over the housewrap. The nailer ensures that the top edge of the wrap won't be caught by the wind and torn off the house.

On roofs with regular rafters, we usually don't use soffit nailers, so we wrap the air barrier over the wall and staple it to the top of the lower top plate. The second top plate holds the housewrap securely.

In winter we often get freezing rain that can encase our framing for days or weeks. When we're concerned that a top plate will become coated, we extend the housewrap up and over the top plate a few inches. After the walls are lifted, we staple the extra wrap inside the wall. We don't cut off the excess until we're ready to set roof rafters or trusses. This approach can save

a lot of time chipping and scraping ice off the top plate. Once the walls are up, we fill in the wing walls, dormers and other areas that are impractical to prewrap. For those hard-to-reach spots when stapling housewrap, we duct-tape our hammer-stapler to a 1x2 furring strip for an extra bit of reach (photo bottom right, p. 119).

Seam tape finishes the job—To get the best performance from housewraps, we seal the seams with seam tape (photo bottom left, p. 119). Seam tape adheres aggressively to housewraps but still needs to be pressed on by hand for best results. We tape what seams we can while walls are down. After the walls are lifted, we tape all the seams we can reach easily right away to lessen wind damage. We leave the second-floor overlap and other high seams for our siding installers to tape while they're on their staging, which is safer than working from a ladder.

Sealing windows and doors is easy when they have nailing flanges. We just tape the flanges (photo facing page). When we install windows with wood casings or brick mold, we first caulk the backside of the window between the casing and the jamb. Next we run a bead of caulk around the opening of the housewrap. We place the unit in the rough opening and press it in place. The caulk seals the unit to the housewrap.

Occasionally, flanged windows are installed before the housewrap goes on. That's no problem; we simply trim the wrap to the edge of the flange and then tape. We also tape any punctures or tears bigger than ½ in.

We've seen some crews lapping housewrap over window and door flanges and over step flashings in an effort to better waterproof their houses. But we never rely on housewrap for this. Most wraps will resist water during construction, but manufacturers note that their products aren't waterproof. Don't use them in place of standard flashing materials. There are materials and details that we use to make our homes more watertight (such as self-adhesive rubberized membranes and copper flashings), particularly those that are near the ocean.

We never bother lining up the 8 in. o. c. marks printed on some air barriers with the framing. We've found the marks to be consistently inconsistent. They commonly vary ½ in. or more in 8 ft. The marks could be useful, but we rarely get a roll that maintains proper spacing.

Be sure to wear sunglasses when working with any of the white wraps. The reflection on a sunny day can be blinding. □

Rick Arnold and Mike Guertin are contributing editors to Fine Homebuilding. *They also are partners in U. S. Building Concepts Inc., a construction-consulting and management business. Photos by the authors, except where noted.*

Skip the housewrap, tape the sheathing?

The whole point of using and properly installing an air barrier is stopping the uncontrolled flow of air into and through a house's wall cavities, and permitting moisture vapor within walls to escape. Because the homes we build are fully sheathed, the only places that air and moisture can enter wall cavities are through the seams between the sheathing, through holes in the sheathing and through the connections between framing and sheathing. With a new tapelike product called Gap Wrap (Benjamin Obdyke Inc., John Fitch Industrial Park, 65 Steamboat Drive, Warminster, Penn. 18974-4889; 800-523-5261), these areas can be sealed. Moisture that enters the wall cavity eventually will make its way out through the sheathing and siding. The issue then becomes whether housewrap is necessary.

A product such as Gap Wrap has some advantages and some limitations. It's cheaper than housewrap and requires less labor to install. The 4-in. tape can be applied by one person, regardless of wind conditions. Some additional caulking may be necessary in areas that don't lend themselves to being sealed with tape, such as wood-to-concrete or wood-to-metal connections, especially in cold weather.

Installing Gap Wrap on oriented strand board (OSB) can be tricky. We've found that the tape does not adhere as well to the coarser surface of some OSB panels as it does to plywood, but the company is working on a glue for OSB.

There are situations where a product such as Gap Wrap won't work. On buildings where siding or T-111 plywood is applied directly to the framing, for instance, properly installed housewrap will create a better air barrier.—*R. A. and M. G.*

Hydronic Radiant-Floor Heating

Concrete slabs, thin slabs and metal plates offer different ways of routing radiant heat underneath every floor in the home

Living in a house with radiant-floor heating can almost make you forget that it's winter outside. I can't think of any other heating system that is as comfortable. Like a campfire on a cool night, heated floors deliver warmth to the skin and clothing without overheating and drying out the surrounding air.

Although usually associated with thick concrete slabs, hydronic radiant systems are now versatile enough to install underneath almost any type of finish flooring. New materials and installation techniques make hydronic heating adaptable to a great range of flooring choices. Although a complete hydronic under-floor heating system requires a boiler, manifolds and controls (to be covered in another article), this article focuses on tubing installation.

Basically, there are three methods of bringing radiant heat underneath the floor of a room: a slab-on-grade system, a thin-slab system and a plate-type system. Each type has advantages and disadvantages, but each system relies on tubing circuits laid underneath the floor. Rooms are heated by warm water circulating through these continuous circuits of tubing. The various circuits receive their heated water from a central supply manifold, a distribution point (like an

Radiant floors can be installed almost anywhere. Radiant tubing is pneumatically stapled to a conventionally framed plywood subfloor before being covered with a thin concrete slab. Flooring choices include wood or carpet.

by John Siegenthaler

Radiant concrete slabs should be well-insulated. Tubing is tied to reinforcing wire mesh and is protected by PVC sleeves where it passes beneath walls and control joints.

electrical junction box) that receives water directly from a boiler. After circulating through the tubing, cooled water returns via the return manifold to the boiler to be reheated.

Special considerations for radiant-floor systems—There are some general rules that apply to all types of hydronic radiant systems that can make the difference between success and failure. First, always have an accurate heating-load estimate for the spaces to be heated. Reputable dealers of floor-heating equipment are a good source of assistance, and most can provide complete designs, including load estimates, layout drawings and help with the sizing and the selection of system components.

Try to include a radiant-floor heating system as early as possible in the house-design process. Accommodations for insulation, load-bearing, floor thickness and other details are more easily and inexpensively resolved early in the process rather than later.

You must be aware that due to their thermal mass, slab-type floor-heating systems are slow to respond to thermostat changes. It can take several hours to bring rooms to comfortable temperatures after prolonged setbacks at lower temperatures, so floor-heating systems are more suitable for locations where stable indoor temperatures are desired.

Don't use floor coverings with high thermal resistance over heated floors. Like wool blankets, high-resistance floor coverings such as plush carpets and pads or multiple layers of plywood under thick hardwood floors effectively insulate the radiant floor from the room that it's supposed to heat. If these floor coverings will be used, make provisions for a supplemental heating device such as a panel radiator or fin-tube baseboard to supplement or even to replace the floor-heating system in that portion of the house.

Tubing circuits should be carefully planned—Because the proper placement of tubing circuits has an impact on the overall performance of a radiant-heating system, I'm a firm believer in drawing a scaled layout of all the tubing circuits in place on the floor plan before beginning installation. This drawing can save hours of trial and error during tubing placement, especially for new installers. Although these drawings can be done by hand, I use a CAD program, ClarisCAD for Macintosh, which saves me considerable time and in some cases even automates the process of determining circuit lengths, floor areas and other repetitive tasks (bottom drawing). And with this diagram in hand, I can draw or spray-paint key portions of the tubing routes right on the insulating foam or floor deck.

I prefer to plan tubing circuits on a room-by-room basis, taking into account at the same time such factors as the design heating load of each room, the necessary tube spacing, the location of the manifold stations and the necessity and location of control joints. Room-by-room planning allows for individual room-temperature control, either when the system is first installed or as an upgrade.

The manifold stations where each tubing circuit begins and ends are usually in interior partitions that allow for an access panel on one side. The inside wall of a closet is a good choice. The manifold station should be located so that tubing circuits to individual rooms go away from it like spokes from the hub of a wheel. This placement minimizes tubing runs down hallways or through other rooms on the way to the intended room.

As a rule of thumb, circuits using ½-in. ID tubing should not be longer than 300 ft. Circuits using ⅝-in. or ¾-in. ID tubing shouldn't be longer than 450 ft. (For more on tubing, see sidebar p. 125.) These values help to limit the pressure drop the system's circulator must operate against as well as the temperature drop that occurs over the course of a circuit. To avoid punctures, it's best to minimize the locations where

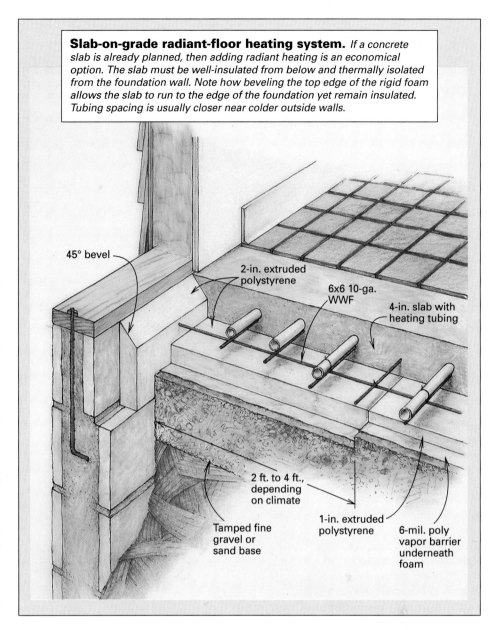

Slab-on-grade radiant-floor heating system. *If a concrete slab is already planned, then adding radiant heating is an economical option. The slab must be well-insulated from below and thermally isolated from the foundation wall. Note how beveling the top edge of the rigid foam allows the slab to run to the edge of the foundation yet remain insulated. Tubing spacing is usually closer near colder outside walls.*

45° bevel

2-in. extruded polystyrene

6x6 10-ga. WWF

4-in. slab with heating tubing

2 ft. to 4 ft., depending on climate

1-in. extruded polystyrene

6-mil. poly vapor barrier underneath foam

Tamped fine gravel or sand base

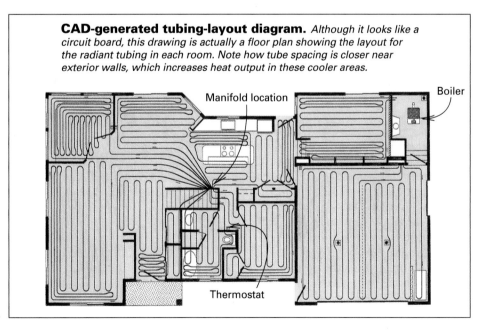

CAD-generated tubing-layout diagram. *Although it looks like a circuit board, this drawing is actually a floor plan showing the layout for the radiant tubing in each room. Note how tube spacing is closer near exterior walls, which increases heat output in these cooler areas.*

Manifold location

Boiler

Thermostat

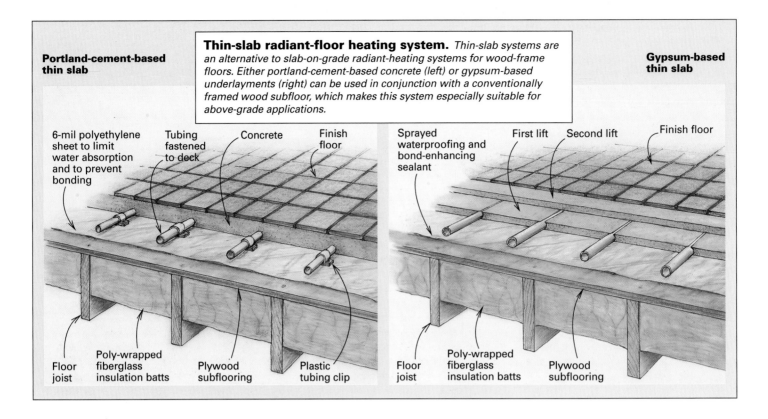

Thin-slab radiant-floor heating system. *Thin-slab systems are an alternative to slab-on-grade radiant-heating systems for wood-frame floors. Either portland-cement-based concrete (left) or gypsum-based underlayments (right) can be used in conjunction with a conventionally framed wood subfloor, which makes this system especially suitable for above-grade applications.*

Portland-cement-based thin slab

6-mil polyethylene sheet to limit water absorption and to prevent bonding

Tubing fastened to deck

Concrete

Finish floor

Floor joist

Poly-wrapped fiberglass insulation batts

Plywood subflooring

Plastic tubing clip

Gypsum-based thin slab

Sprayed waterproofing and bond-enhancing sealant

First lift

Second lift

Finish floor

Floor joist

Poly-wrapped fiberglass insulation batts

Plywood subflooring

tubing passes under partitions that will be mechanically fastened to the slab. If this can't be avoided, mark the locations of buried tubing and make sure to use short fasteners. If the framer is agreeable to fastening the partitions with construction adhesive, a method that works well but that requires a warm, dry slab when the adhesive is applied, the tubing can be run under partitions without restriction.

Also, it's best to minimize the locations where tubing crosses under sawn control joints in the slab. This reduces the number of control-joint sleeves as well as the possibility of nicking a tube that might have been improperly lifted under a control joint. Leaks are repairable with special fittings, but the process is both time-consuming and costly.

Slab-on-grade systems are cost-effective—A hydronic floor-heating system installed as part of a slab-on-grade floor is the most economical type of radiant system, and the most common. Because the installed concrete slab is already factored into construction costs, the extra cost for floor heating is essentially that of installing the tubing and subslab insulation.

Adequate insulation is an essential component of any slab-on-grade system (top drawing, p. 123). Extruded-polystyrene insulation both under the slab and near its exposed edges reduces downward and edgewise heat loss, which would otherwise waste a large portion of the energy supplied to the slab. Prior to installing this insulation, all under-slab piping and wiring

Gypsum-based thin slabs are generally installed by factory-trained crews. These underlayments are self-leveling and strong, but they are susceptible to moisture and abrasion without an appropriate finish floor.

should be in place. The entire area should then be leveled, thoroughly tamped and covered with a 6-mil poly vapor barrier generously lapped at any seams. Two-in. extruded polystyrene sheets placed vertically between the edge of the slab and the foundation wall ther-

mally isolate the slab and can be held in place temporarily with foam-compatible construction adhesive. Ripping a 45° bevel along the top edge of the foam before installing it allows the concrete slab to cover the exposed edge of the insulation and to provide a smooth base for the finish flooring yet still remain thermally isolated from the exterior wall.

A 2-ft. to 4-ft. width (depending on climate) of 2-in. thick extruded polystyrene should also be placed underneath the slab around the perimeter. The remaining interior portions of the slab should have a minimum of 1 in. of subslab insulation except where structural bearing pads (footings inside the perimeter walls) are located. And experience has taught me that it's a good idea to place welded-wire mesh reinforcement and spare planks or blocks over the foam as it is being placed. Otherwise, a good breeze can lift everything up and blow it away.

Of course, different foundation-wall designs, varying climate conditions and local-code requirements may necessitate other insulation details. The idea is to provide adequate insulation for the climate and a continuous thermal break between the heated slab and the colder surrounding materials.

Tubing is anchored to wire-mesh reinforcement—I establish the eventual location of the manifold station before actually laying down any tubing. A predrilled template block with two rows of holes spaced 2 in. o. c. can be used to anchor tubing in the correct location and orientation during the pour. This template should be exactly in line with the partition where the manifold station will be located, so carefully measure its exact location relative to the foundation walls.

Begin each circuit by pushing 2 ft. to 3 ft. of tubing up through one hole in the block; then unroll the tubing from its coil and progressively fasten it down along the length of the circuit. End the circuit by pushing another 2 ft. to 3 ft. of tubing up through the hole in front of or behind the starting hole.

The tubing is secured to the welded-wire reinforcing mesh using either soft wire ties, plastic pull ties or plastic clips. Plastic ties and clips are endorsed by more tubing manufacturers than are metal ties, and in some cases tubing warranties are contingent on the use of approved fastening methods. Ties are located about every 24 in. to 30 in. on straight runs with three ties at each return bend. Ties hold tubing in place and prevent it from floating up during the pour, so they don't need to be overly tight. Care should be taken that the loose ends of the ties aren't left pointing up where they could interfere with concrete finishing.

Tube spacing depends on several factors, including the rate that heat must be released from the floor, average water temperature of the heating system and the type of floor covering installed over the slab. Typical spacings for slab-on-grade jobs are 6 in., 12 in. and 18 in. o. c., which correspond nicely with the 6-in. grid pattern of welded-wire reinforcing mesh. Closer 6-in. spacing is common near the edges of the slab where heat losses are greater, or in areas with higher-thermal-resistance floor coverings, such as carpeting. Twelve-in. spacing is common for many residential applications in buildings with average heating loads and average areas of glass. However, 12-in. spacing should be limited to rooms where the upward-heat-flow rate does not need to exceed 20 Btu per sq. ft. per hr. Interior rooms with relatively low heating loads can sometimes get by with 18-in. spacing.

Other spacings, though possible, can complicate fastening the tubing to standard reinforcing wire. Combinations of spacing can sometimes be used within the same room. For example, the beginning of the circuit may use 6 in. o. c. spacing near an exterior wall with a large area of windows, and then make a transition to 12 in. o. c. spacing after it has progressed 3 ft. to 4 ft. in from the wall.

After all the circuits are in place, they must be pressure-tested to at least 60 psi for 24 hours to verify that no leaks are present. Compressed air or water can be used for this test, although water shouldn't be used in freezing temperatures. The tubing should remain under pressure during the pour. If necessary, wheelbarrows full of concrete

Tubing choices. Left to right: Vanguard polybutylene tubing; IPEX's Kitec PEX-AL-PEX tubing; Heatway's Entran synthetic rubber tubing with oxygen-diffusion barrier; Wirsbo PEX tubing with oxygen-diffusion barrier; and Maxxon's Infloor PEX tubing (without oxygen-diffusion barrier).

Understanding hydronic-tubing options

Although copper tubing was once extensively used for radiant floors, its current cost, short coil length and tendency to expand and contract at different rates than the slab it's embedded in make it a poor choice compared with other tubing now available. Nowadays, nearly all floor-heating systems installed in North America use either polymer (plastic) or synthetic-rubber tubing. The three types of polymer tubing most often used are cross-linked polyethylene, or PEX; an aluminum/PEX composite tube often called PEX-AL-PEX; and polybutylene (photo above).

PEX tubing has a proven track record— Of the three, PEX holds the greatest share of the floor-heating market worldwide, and it's my first choice when I specify tubing. Wirsbo (800-321-4739) estimates that it has over 2 billion ft. of PEX in service.

PEX tubing has a "shape memory" characteristic that allows accidental kinks to be repaired on the job. Heating the kinked area with a heat gun to approximately 275°F will make the kink disappear without permanent damage to the tubing.

IPEX (800-473-9808) manufactures a composite tubing called Kitec, which consists of a thin-wall aluminum core with layers of PEX bonded both inside and outside. Because it's comparable in cost to PEX and has been used successfully in Europe and Canada, look for this type of tubing to gain a major market share in the United States in the near future.

Unlike PEX tubing, PEX-AL-PEX's aluminum core enables it to retain its shape when bent. Return bends in floor circuits can be easily hand-bent to a minimum radius equal to five pipe diameters. The aluminum core also provides a nearly perfect oxygen-diffusion barrier (more on this later).

Polybutylene tubing is also used for floor heating. Polybutylene tubing is not cross-linked, and kinks can't be field-repaired, except by cutting out the section and splicing in a coupling. The current cost of Vanguard's (316-241-6369) polybutylene tubing (with an oxygen barrier) is comparable to that of PEX tubing. Unfortunately (and for reasons

unrelated to its performance in floor-heating applications), polybutylene tubing will no longer be available in the United States by early 1997. Look for polybutylene suppliers to begin distributing PEX tubing to be used as a replacement for radiant floors.

Synthetic-rubber tubing, such as Heatway's Entran (800-255-1996), is also manufactured in the United States specifically for hydronic floor-heating applications. Consisting of several layers of synthetic-rubber compounds and a reinforcing mesh, it is flexible and highly resistant to kinking damage. The cost of synthetic-rubber tubing is comparable to that of PEX and polybutylene tubing.

Oxygen-diffusion barriers are important—A characteristic of both polymer and synthetic-rubber tubings is their inability to stop oxygen molecules from diffusing through their walls. This steady flow of oxygen into the system can severely corrode iron or steel components, such as boilers and circulators, and will almost certainly lead to premature failure. Fortunately, this deficiency can be corrected with the addition of an oxygen-diffusion barrier to the tubing. Nearly all manufacturers of polymer floor-heating tubing, be it PEX, PEX-AL-PEX, polybutylene or synthetic rubber, currently offer tubing with such a barrier to reduce oxygen diffusion to an insignificant level. Tubing that is equipped with an oxygen-diffusion barrier is a must for systems with cast-iron components or steel components.

Most tubing intended for floor-heating application is available with nominal inside diameters from 3/8 in. to 3/4 in. For short residential circuits or for places where close tube spacing is needed, 3/8-in. or 1/2-in. tubing is suitable. To limit pumping pressure, circuits with 3/8-in. tubing should not be more than 200 ft. long. Those with 1/2-in. tubing should not exceed 300 ft. For both residential and commercial systems, 5/8-in. and 3/4-in. tubing can be used. Circuits with 5/8-in. tubing should not be more than 450 ft. long. If properly designed, circuits using 3/4-in. tubing can be upward of 500 ft. long.—*J. S.*

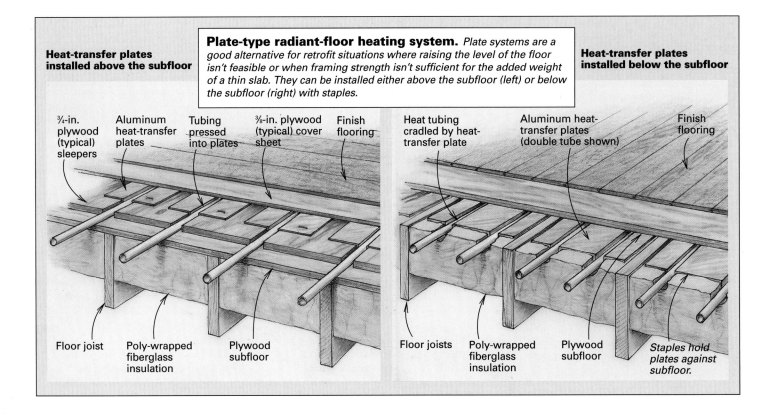

Plate-type radiant-floor heating system. *Plate systems are a good alternative for retrofit situations where raising the level of the floor isn't feasible or when framing strength isn't sufficient for the added weight of a thin slab. They can be installed either above the subfloor (left) or below the subfloor (right) with staples.*

Heat-transfer plates installed above the subfloor

¾-in. plywood (typical) sleepers

Aluminum heat-transfer plates

Tubing pressed into plates

⅜-in. plywood (typical) cover sheet

Finish flooring

Floor joist

Poly-wrapped fiberglass insulation

Plywood subfloor

Heat-transfer plates installed below the subfloor

Heat tubing cradled by heat-transfer plate

Aluminum heat-transfer plates (double tube shown)

Finish flooring

Floor joists

Poly-wrapped fiberglass insulation

Plywood subfloor

Staples hold plates against subfloor.

can be rolled over the tubing without damaging it. Take special care not to pinch tubing under the metal nose bar as the wheelbarrow is dumped, though.

The wire reinforcing with the attached tubing should be supported during the pour with wire or plastic high chairs, or lifted while the concrete is being placed, so that the tubing is about halfway through the slab's thickness, except where sawn control joints will be made. At these locations tubing and reinforcing wire should remain near the bottom of the slab, and the tubing should be protected with a sleeve made of polyethylene or PVC tubing (bottom photo, p. 122).

Installed costs for slab-on-grade systems in my area (upstate New York) run about $1.75 to $2 per sq. ft. of heated floor. This includes ⅝-in. PEX tubing 12 in. o. c., 1-in. extruded-polystyrene insulation under interior slab areas (2 in. near slab edges), installation labor for the tubing and an allowance for manifolds.

Thin-slab systems can offer more installation versatility—Hydronic floor heating can also be used in combination with wood-frame floors. A common approach is to fasten tubing to subflooring, and then to cover it with a thin 1¼-in. to 1½-in. layer of poured underlayment (drawing p. 124).

One material that's become popular for thin-slab systems is poured gypsum-based underlayment, such as Maxxon Corporation's Thermafloor (920 Hamel Road, Hamel, Minn. 55340-9610; 800-356-7887). This product, a blend

of gypsum cement, additives, masonry sand and water, is sold only as an installed product, meaning you can't buy it and install it yourself.

Factory-authorized crews proportion and mix the ingredients in a portable rig set up outside the house, pumping the resulting slurry into the house through a hose. With the consistency of a milk shake, it is essentially self-leveling as it flows onto the floor (photo p. 124).

Because gypsum-based underlayments shrink as they dry, they should be poured in two separate layers, or lifts, when used in floor-heating applications. The first lift is poured even with the top of the tubing and is usually firm enough to walk on after about two hours. Shrinkage will leave the top of the tubing protruding above the first lift, so the second lift easily covers this slight irregularity and produces a smooth surface.

Although the final compressive strength of gypsum-based underlayments ranges from 1,500 psi to 2,500 psi, they are too susceptible to moisture and abrasion to act as a finished wearing surface. Finish flooring such as sheet vinyl, ceramic tile, carpeting or laminated strip wood flooring are all suitable for gypsum-based underlayments, but it is crucial that the slab be thoroughly dry, which typically takes at least a week, before any of them is applied. In some cases the slab must also be coated with a specific waterproofing and bond-enhancing sealant. Always verify that the finish-flooring material will be warranted over a heated gypsum-based underlayment, and carefully follow the written installation specifications.

Concrete thin slabs need control joints—An alternative to poured gypsum-based underlayments is portland-cement-based concrete made with fine (#1A) pea-stone aggregate. Admixtures such as water reducing agents, superplasticizers and Fibermesh are often added to the mix to increase slump, improve strength and reduce shrinkage.

Installing concrete thin slabs is easiest when it's done before any partition walls are in place. The wall locations and the tubing-circuit locations can be laid out on the deck, and then the deck is covered with 6-mil clear-polyethylene film. The poly acts as a slip sheet and prevents the concrete from bonding to the deck, allowing for differential movement between the slab and the framing. Otherwise, stresses can develop that could randomly crack the slab.

Two-by plates are then fastened over the poly to the layout lines. This creates a 1½-in. deep pan for the concrete on the deck. Toilet flanges and any other mechanicals should also be blocked up 1½ in. Coating the edges of all of the plates with mineral oil or another form-release agent will prevent the concrete from bonding to the plates.

The slab should be segmented into small sections that more easily absorb seasonal movement of the floor framing without random cracking. This is done with control joints, created by stapling the 1-in. PVC angles that are typically used for drywall-corner protection (Trim-Tex, 3700 W. Pratt Ave., Lincolnwood, Ill. 60645; 800-874-2333) to the subfloor before the pour. These strips create fault lines in the slab and control the inevitable cracking. I typically place these control joints at doorways and inside corners, and I segment large rooms into smaller sections with them.

Concrete can be placed either from a wheelbarrow or directly from a concrete-truck chute. The 2x plates are perfect for screeding, and the slab is then floated and troweled as required for the selected finish floor.

Both gypsum-based and portland-cement-based slabs have pros and cons. Although the gypsum-based products are arguably faster and easier to install (by a trained crew), a concrete thin slab has higher thermal conductivity, excellent resistance to moisture and often significantly lower cost. The final decision about which to use may rest in the cost and availability of each system in a region or at a job site.

Thin-slab systems using gypsum-based underlayments run between $4.25 and $4.50 per sq. ft. of heated floor in my area. Those using pea-stone concrete have run between $2.25 and $2.75 per sq. ft. Both costs include ⅝-in. PEX tubing 12 in. o. c., slab materials, underside insulation, allowance for manifolds and labor.

Radiant systems don't necessarily need thermal mass. Heat-transfer plates conduct heat away from tubing and into the floor system. These plates can be installed either above or below the subfloor, which makes them a good choice for retrofit installation.

Accommodating a thin-slab system—Regardless of which material is used, several factors must be considered in preparing for a thin-slab installation. For example, the floor framing must be able to support an additional 15 lb. to 18 lb. per sq. ft. of dead-loading due to the added weight of the slab. Although this additional framing strength can easily be planned for in new construction, the added cost of beefing up deficient framing in an existing building usually rules out this option.

Also, the heights of window and door rough openings, base cabinets, stair risers and toilet flanges must be raised to accommodate the added thickness of the slab. Again, this accommodation is easier to accomplish in new construction than in retrofits.

In any kind of thin-slab radiant installation, the underside of the floor must be insulated to limit downward heat loss. I use R-11 insulation for floors above heated space, R-19 for floors above semiheated basements and at least R-30 for floors over vented crawlspaces. I like to use poly-wrapped fiberglass batts for this, though other options include rigid foil-faced foam cut to fit or flexible foil-faced batts.

Plate-type systems are good for retrofits—Plate-type systems rely on aluminum heat-transfer plates that wrap partially around tubing and conduct heat away and into the floor system (photo above). The tubing and plates can be installed either above (on sleepers) or below plywood subflooring, the latter being more

common because of its lower cost and faster installation. Plate-type systems are especially well-suited for retrofit applications because they don't disturb existing flooring and add little weight to the floor (drawing facing page).

Because there is less contact area between the plates and the tubing as compared with tubing that is fully embedded in a slab, tube spacing is typically no more than 8 in. o. c. In the majority of cases, the supply-water temperature must also be higher for plate-type systems than it is for slab-type systems.

Tubing is routed down the space between the joists, makes a U-turn at the end and comes back up the same cavity. The tubing required for each joist cavity must be pulled through holes drilled near one end of the joists, which should be at least ½ in. or so larger than the outside diameter of the tubing. This makes it easier to pull the tubing through. Keep in mind that holes drilled in solid-wood floor joists need to be placed a minimum of 2 in. from either the top or the bottom of the joist or to another hole in the joist, and that the maximum size of these holes cannot be more than one-third of the joists' depth.

Heat-transfer plates, which cradle the tubing, are then stapled to the underside of the subflooring. Plates that allow the tubing to be snapped in place after the plate has been fastened to the subflooring are also available.

Both tubing and plates expand when heated and contract as they cool, and sloppy installation is certain to cause expansion/contraction noises as the system operates. To avoid these

noises, make sure that the tubing enters and exits the plates in alignment with the plates' centerline. Bends in the tubing should be as gentle as possible, and holes where the tubing passes through the joists should be oversize at least ½ in. to prevent any binding. A minumum ¼-in. expansion gap should be included between adjacent plates.

Plate-type systems can't be seen from the top of the floor deck, so be sure all trades on the job are aware of their presence beneath the floor before a nail, a sawblade or a drill bit makes the discovery for them.

In assessing retrofit jobs for the possibility of a plate-type system, be sure to examine the underside of the floor decking. Interference from flooring nails, plumbing, wiring, bridging and any other types of obstacles can significantly slow installation or even eliminate the system from consideration.

Plate-type systems tend to be more expensive than thin-slab systems because of the closer tube spacing and the extra cost of plates. I currently use an installed cost of $5.50 per sq. ft. of heated space for estimating purposes. This includes ⅝-in. PEX tubing 8 in. o. c., plates, underside insulation, manifold allowance and labor. □

John Siegenthaler, P. E., is the author of Modern Hydronic Heating *(Delmar Publishers, 1995), a consulting engineer and an associate professor of engineering technology at Mohawk Valley Community College in Utica, New York. Photos by Andrew Wormer, except where noted.*

Mixing Forced-Air and Boiler Heat

This system provides air conditioning, radiant-floor heat and domestic hot water

by Richard D. Groff

The installation of heat pumps is virtually automatic in houses built in my part of the country, Pennsylvania's Susquehanna Valley. So I feel fortunate when my family's plumbing and heating company gets the chance to design and install a heating system that strays from this norm. Recently, we got just such a challenge when we were invited to bid on the heating system for a new 3,200-sq. ft. house.

The homeowner didn't plan on moving again, so he put a priority on staying comfortable year-round, even if that meant spending more for a heating system than he otherwise would. The homeowner understood heating and cooling design and had settled on a hydronic heating system, which is a system that uses hot water as the medium to heat the house. But he didn't want the source of heat—the baseboard units installed with most hydronic systems—to be visible. Nor was that his only requirement. The client also wanted central air conditioning, a means of filtering and humidifying air for the house, radiant heat in the master-bathroom floor and enough domestic hot water to fill a large whirlpool bath without overloading the system.

That's a long list of requirements for any heating system and far exceeds what a standard heat-pump system would be able to accomplish. The heating system we devised, however, met all of the owner's needs by combining advantages found in both forced-air and hot-water systems. The initial cost was higher than it would have been for a heat pump, but the heating system will last longer and perform more efficiently.

Why not a heat pump?—In designing any heating and cooling system, we start with the temperature parameters appropriate in central Pennsylvania—and that helps explain why a heat pump would not have been the best choice. We design a system so that it is capable of maintaining a 70° F indoor heat level when the outdoor winter ambient temperature is 0° F. In summer the cooling system should be able

Many homeowners don't want a heat pump. The heat isn't even, and it's expensive to give the system the quick boost it needs in real cold weather.

High-volume water heater. **Domestic hot water could have been heated with a coil installed directly in the boiler, but for a higher flow rate, the author used an indirect-fired water heater (left) instead.**

to maintain an indoor temperature of 75° F when the outside temperature is 95° F.

One reason why heat pumps are so popular is that the same unit heats and cools a house. In winter a heat pump extracts latent heat from outside air (some heat can be removed from air at all temperatures above absolute zero). During the cooling season, a heat pump reverses the operation. Instead of removing heat from the outside air, the pump removes heat from indoor air, thus cooling and dehumidifying air in the house. The trouble is that heat pumps lose operational efficiency at temperatures below 30° F. Only when the outdoor ambient temperature is between 30° and 35° can the heat pump maintain 70° indoors—that's called the system's balance point. When the temperature drops below the balance point, the heat pump runs constantly while steadily losing ground on the desired indoor air temperature of 70°.

To make up the difference, we commonly add electrical-resistance heat elements to the air-distribution system. For a better idea of how this works, picture yourself driving a car down a long grade and then starting up a hill. You wouldn't need the accelerator until the downhill momentum (your heat output) is exhausted. The car gradually would come to a place on the hill where all the downhill momentum is gone (the balance point), so you press the accelerator (the electrical-resistance heat) to get the car to the top of the hill. That's why this homeowner, and many others we serve, didn't want a heat pump: The heat isn't even, and it's expensive to give the system the boost it needs in real cold weather. The supply air in a heat-pump system, at 95° or 100°, cools quickly as it moves across the floor. So when the air contacts you, it can be rather uncomfortable. We can make the supply air warmer with electrical-resistance heat—and pay the cost—or look for more effective heating options.

Our answer is a hybrid system— The shortcomings of a heat-pump system convinced the homeowner

Coils for cooling. The air-conditioning coil on the floor will be mounted inside the cabinet, which also contains the blower. The system uses two such cabinets, or air handlers, to provide heat and cooling.

Coils for heating. The author installs a hot-water coil on top of an air handler. The coil, drawing hot water from the boiler, heats supply air before it is pumped into the ductwork.

Electronic air cleaners. The electronic air-cleaning units are efficient without impeding the flow of air through the ducts. Plates that collect particles from the air can be washed off.

that the best option would be hydronic heat. The source of the hot water is an oil-fired cast-iron boiler (often called a furnace by mistake); the hot water produced in the boiler is the medium for heating the house. In standard hydronic systems, hot water is pumped from the boiler through fin-tube units along the baseboards (or through cast-iron radiators) in each room. The fin-tube radiators warm the air, and the water is returned to the boiler to be reheated. Even though the homeowner wanted hydronic heat, he didn't want to look at any baseboard heat units. So I decided to heat the house with warm air delivered through ductwork and registers, just like any forced hot-air heating system. The difference is that the air is warmed by hot-water coils, not heated directly in a furnace. The ducts used to distribute heat around the house double as air-conditioning ducts in the summer.

One advantage of this approach is that it allowed us to use separate heating and cooling units, which are more efficient than a heat pump doing both jobs. Another plus is that the boiler also can be used to heat water for radiant floors in the master bathroom. Radiant heat involves heating water and pumping it through flexible plastic tubes installed beneath the floor. Radiant heat warms just about any kind of floor—wood, carpet, vinyl and especially tile and stone. An entire floor surface serves as the heating element, or source of heat.

A hydronic system also provides domestic hot water. This is usually achieved by installing a hot-water coil directly in a boiler. These coils effectively provide hot water for household use, but the flow rate can be inadequate for heating large volumes of water. Our clients had chosen a 100-gal. whirlpool bath, and we anticipated abnormally high demands for domestic hot water. With that in mind, we picked an indirect-fired, domestic water heater that can provide up to 135 gal. of hot water per hour (photo p. 128). In an indirect-fired water heater, like the Amtrol heater we used (Amtrol, Inc., 1400 Division Road, West Warwick, R. I. 02893; 401-884-6300), hot water is piped in from the boiler to keep the 30-gal. reservoir hot—just the opposite of a coil installed directly in the boiler. The advantage of this system is that it provides much higher flow rates than a typical domestic hot-water coil does.

We certainly don't install a high-tech heating system like this every day. The system, at about $13,500, was almost twice as costly as a standard heat pump would have been. But comfort will be unparalleled, and the boiler should last for

Warming the floor. Starting with flat aluminum stock, the author formed heat-distribution plates that snap over the flexible polyethylene tubing. The plates are stapled to the plywood subfloor and are heated by hot water flowing through the tubing.

30 years. The cooling units should last twice as long as a heat pump would. In the long run, the system is quite appealing.

Heating and cooling as separate entities— Although warm and cool air are both routed through the same ducts and driven by the same blowers, the cooling and heating units are separate. We calculated the cooling load to be 4 tons (a ton of cooling capacity is equal to 12,000 British thermal units, or Btu). The cooling load is divided equally between the first and second floors of the house, so we decided to use dual 2-ton air-conditioning systems (one for each floor), each controlled separately, instead of a single 4-ton system.

For efficiency, we decided to install both air-handling units and the ductwork trunk lines in

the basement (air handlers are the cabinets that contain system blowers, cooling evaporative coils and controls). In the basement the temperature is constant and not subject to the severe hot or cold swings that wreak havoc when these systems are installed in attics (attic installations can rob a system of up to 30% of the conditioned air they have just created). In this house, basic heating and cooling principles dictated that we move at least 400 cu. ft. per minute (cfm) per ton of cooling, so we specified two 1,000-cfm air handlers (top left photo, p. 129). Two, 2-ton high-efficiency cooling coils were installed and matched with condensers to complete the cooling system.

On the heat side, we chose a four-section, cast-iron boiler sized to meet the load requirement of the house. We use only cast-iron Burnham boilers, which are made just up the road (The Burnham Corp., P. O. Box 3079, Lancaster, Pa. 17604; 717-397-4701). The boiler runs at an efficiency of nearly 84%. We designed the system so that hot-water coils were mounted on top of each air handler (bottom left photo, p. 129). When a thermostat calls for heat, hot water is circulated through these coils, warming the supply air before it is pushed through the ducts and into the house.

The trunk lines in the basement are made from duct board, 1-in. thick compressed fiberglass that we buy in 4x10 sheets and form into ducts. When we fabricate ducts, we make them right on the job and get exactly what we want. We are careful not to expose any raw edges of the duct board to the airstream so that fiberglass particles won't get into the supply air. Smaller ducts serving individual rooms are made of sheet metal. The duct system delivers the air mostly in a vertical pattern in each room, from the floor toward the ceiling. This air-delivery pattern helps prevent air stratification (thermal layering of air). Sidewall registers were used sparingly and only where absolutely necessary. And we never deliver air from a ceiling when heating because warm air will hug the ceiling while floor areas remain cool. This is yet another reason to stay away from attic installations—supply-air registers invariably end up in the ceiling.

Both high and low air-return registers were used (about 1 ft. from the floor and 1 ft. from the ceiling) in all rooms except the kitchen, the bath and the laundry where odors might be present. Closing off the low air returns in the summer allows the homeowner to pull unwanted warm air from the ceiling. The air is cooled and dehumidified by the air-conditioning system. During the heating season, opening lower returns and closing those near the ceiling reverses the effect: Uncomfortably cool air near the floor can be removed.

Maintaining indoor air quality—
Just before the return air enters the air handlers, it passes through Honeywell electronic air cleaners (right photo, p. 129) (Honeywell, Inc., Honeywell Plaza, P. O. Box 524, MN27-2164, Minneapolis, Minn. 55440-0524; 612-951-1000). We installed one in each of the two air handlers. These units remove almost all dust, pollen and smoke from the air by electrically charging the pollutants. The particles are attracted to collector plates, which can be washed off every month or two. Unlike electrostatic filters, the Honeywell electronic filters cause almost no air resistance. Electrostatic filters do almost the same job at a lower cost, but if neglected by the homeowner, a dirty electrostatic cleaner may damage heating or cooling equipment by impeding air flow.

Another issue to tackle was humidity, especially during the winter. Repeated recirculation of air within a living space removes some of its moisture content. The warmer the air, the more prevalent the drying effect. Humidity levels of 40% or lower can be expected if this issue goes unresolved, and air that dry nags at noses, throats and eyes. For some people, it's downright unbearable. In addition, studies show that people are satisfied with lower temperature settings in winter if the air is not too dry. By lowering the thermostat setting a few pegs, fuel costs can be lessened.

We installed a humidifier in each of the two air handlers. A rotating drum of spongelike material is spun through a small reservoir of water at a slow speed. Warm supply air passes through the drum, picks up moisture and is fed into the return airstream. A humidity-sensing device allows adjustment of the amount of water introduced into the supply air. With the supply air at 140° F, each of these units can put as much as 22.5 gal. of water into the house's living spaces every day.

Radiant heat in the bathroom—The homeowner dreamed of having a tile floor in the large, second-floor master bathroom. He wanted barefoot comfort year-round. But no matter how warm the supply air, a tile floor feels cool under foot during the winter months—even if the floor is located above a heated room. This problem is solved by heating the floor itself with radiant-floor heat. This technique has come a long way,

Copper-to-plastic fitting. **To make the connection between flexible plastic tubing and the copper supply lines, a barbed connector is used. The end of the plastic tube fits over the end of the connector; the joint then gets another plastic sleeve and a stainless-steel hose clamp.**

but the basic difference between past and present radiant-heat systems is the piping. State-of-the-art plastic tubing is tremendously flexible and essentially indestructible.

One way of distributing heat in a radiant-floor system is to embed flexible tubes in lightweight concrete and install the finish floor over the concrete. But radiant tubing also can be applied under a plywood subfloor, between floor joists, and that's the approach we decided to go with. Because we had to provide heat through a plywood subfloor, cement board and then ceramic tile, we wanted a material with excellent heat-transfer characteristics. We used aluminum about 1/16 in. thick from The Radiantec Company (P. O. Box 1111, Lyndonville, Vt. 05851; 800-451-7593) that we fashioned into heat-distribution plates. The plates are 10 in. by 16 in. and are de-

signed to pull warmth from the piping and distribute it to a larger surface area so that the floor is heated evenly. We took the flat stock that we bought from Radiantec and formed plates that snap right around the heat-exchange tubing. The plates are stapled to the underside of the subfloor with 1/4-in. or 3/8-in. staples. We kept the plates 8 in. apart and used about 16 staples per plate (photo facing page).

Radiantec also provided the thin-wall heat tubing that we use, a type of polyethylene that is designed to control the amount of oxygen in the hot-water loop so that heating water is less corrosive on the boiler. Walls of the 3/4-in. I. D. tubing are 1/16 in. thick.

To provide heat in the bathroom above, we calculated that we would need one run of tubing in each joist bay. After running the tubing down one bay, we drilled through the joist about 2 in. from the top and 6 in. from the end of the bay and routed the tubing into the next bay. In this fashion we zigzagged the tubing beneath the entire floor area.

We used ribbed fittings from Radiantec to connect the plastic tubing to 3/4-in. copper supply lines for hot water (photos left). Then we installed 3 1/2-in. foil-faced fiberglass insulation (with the foil facing up) between the joists where the tubing was located. The insulation reflects heat upward. Because the bathroom has two heat sources—forced hot air and the radiant floor—we didn't want to use a conventional air-sensing thermostat to control the hot-water circulator under the floor. Instead, we installed a Honeywell T675A control with a remote sensor to prevent the radiant floor from cycling on and off because of fluctuations in air temperature. The Honeywell sensor is installed inside the vanity to buffer it from drafts. In this way the system is able to maintain an even floor temperature no matter what the ambient air temperature is in the room.

To make the floor comfortable under bare feet, we aim for supply water of 110° F in the tubing, and a drop in water temperature from the supply to return side of no more than 20° F. A mixing valve tempers the 170°-to-190° boiler water to the desired 110°. To prevent overheating, it's a good idea to install a thermometer and limit controller directly in the hot-water line to shut off the circulator in case the mixing valve malfunctions. It should be wired to shut off the circulator at no higher than 145°. □

Richard D. Groff is in charge of heating and air conditioning for Neffsville Plumbing & Heating Services in Neffsville, Pa. Photos by Lin Wagner.

Framing Corners

A survey of methods and materials used to build warm, inexpensive and strong corners

by Charles Bickford

Unless you are a builder who always builds round houses, chances are pretty good that a portion of your building career will be spent framing corners. Until recently, framing a corner was a straightforward task that didn't present too many options. The typical stick-framed corner was, like its predecessor the timber post, a massive piece of wood. It was sturdy, and you could drive nails into it anywhere you pleased. These days, the traditional three-stud corner is not gone but may soon become as rare as solid-chestnut timbers. For a variety of reasons, there is now more to consider than just providing nailing support for siding outside and drywall inside.

The primary reason is economics. The price of framing lumber has nearly doubled in the past ten years. The prevailing opinion holds that the lumber's quality has slipped considerably, too, which is an economic factor if you have to spend 20 minutes digging in the pile trying to find a straight 2x for a sill or a plate. The cost of heating a house has also changed the way people frame corners. During the energy crunch of the 1970s, some architects and builders be-gan to think harder about more efficient ways to insulate houses and switched from 2x4 to 2x6 walls for their greater R-value potential. Corners, always noto-rious cold spots, came under scrutiny, too. If built with 2x6s, the older style sol-id corners not only required more lumber than the 2x4 version, but they also created a bigger area in the wall that couldn't be insulated and suffered more from thermal bridging. Thermal bridging is the transfer of heat energy through solid materials; in a house, cold is usually conducted from the outside sheath-ing to the interior walls through any solid material, most notably the framing.

In the interest of both economy and energy efficiency, researchers and builders have been searching for better ways of framing corners using less wood and more insulation. Here are some of the methods that they've come up with. □

Charles Bickford is an assistant editor at Fine Homebuilding. *Photos by Scott Phillips, except where noted.*

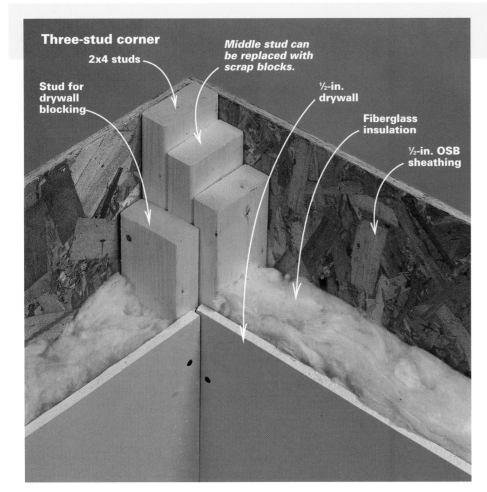

Three-stud corner

2x4 studs

Middle stud can be replaced with scrap blocks.

Stud for drywall blocking

½-in. drywall

Fiberglass insulation

½-in. OSB sheathing

The three-stud corner is traditional and sturdy

The three-stud corner (photo left) is a direct descendant of the timber post and is fa-vored by many builders. Its main advan-tages are that it is quick to assemble, it is strong and it gives the carpenter plenty of support for nailing exterior corner boards and siding. You can make it two ways: Nail three full studs together, or use 2x scrap from the site as blocks between two full studs. A single end stud on the abutting wall is nailed on the inside corner. The main disadvantages of three-stud corners are that they use more lumber, are difficult to insulate and create a wide thermal bridge to the interior. These corners are not effective when built with 2x6s. Connecticut framer Mario Sapia uses this corner be-cause he like the mass of the construction; he reduces the thermal bridging by wrap-ping the exterior of the house in ¾-in. rigid-foam insulation.

Sidebar photos, facing page: Charles Bickford

Drywall clips: engineered hardware that takes the place of wood

Carpenters and wood are as inseparable as bakers and flour. But unlike wheat, trees cannot be harvested on an annual schedule, and we all know that lumber is not as plentiful, cheap or straight as it used to be. One way to slow the drain on your wallet and on lumber supplies is to use drywall clips.

Clips are used in place of wood backing to support drywall in corners (walls and ceilings). Depending on the style, drywall clips are slipped on to the drywall or nailed to the stud and can be installed either by the framers or by the drywall crews. All clips are usually installed 16 in. o. c. to support the first sheet of drywall in the corner; any protruding tabs are covered by the succeeding sheet. Because the clips take the place of a stud in a corner, there is also greater space in the wall cavity to fill with insulation.

The clips certainly are not mainstream items, even though they've been on the market in some form for more than two decades. Contractors familiar with the product say that the clips offer a sturdy and economic alternative to lumber. The clips run from 12¢ to about 20¢ each. These clips may be hard to find at the lumberyard, so check the Yellow Pages under "Drywall Supplies" or contact manufacturers. Included here is a partial listing of drywall-clip manufacturers and is by no means inclusive.—C. B.

Sources of supply

DS Drywall Stop
(light-gauge steel)

Drywall Stop
Simpson Strong-Tie Co.
4637 Chabot Drive, #200
Pleasanton, CA 94588
(800) 999-5099
http://www.strongtie.com

The Nailer
(recycled HDPE)

The Nailer
The Millennium Group
121 S. Monroe St.
Waterloo, WI 53594-1407
(800) 280-2304

Corner-Back fasteners
(light-gauge steel)

Prest-on Clips
Prest-on Co.
312 Lookout Point
Hot Springs, AR 71913
(800) 323-1813

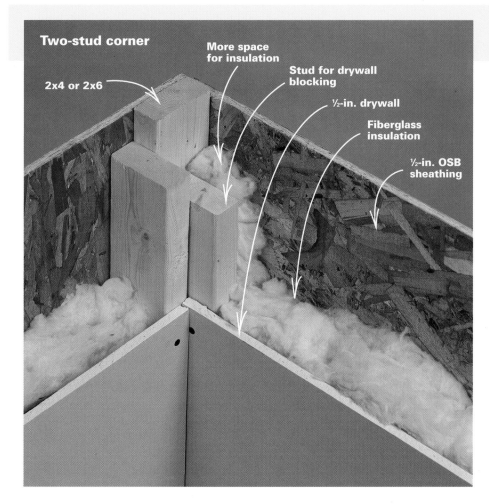

Two-stud corner

2x4 or 2x6

More space for insulation

Stud for drywall blocking

½-in. drywall

Fiberglass insulation

½-in. OSB sheathing

The two-stud corner uses less wood and is easily insulated

The two-stud corner (photo left), also known as the California corner, is popular with builders because it goes together quickly and because it uses less lumber. It is also more energy efficient because it has less mass to create a thermal bridge and opens a larger space that's easy to insulate. Some builders don't use it because the corner gives them less to nail to if they're putting up vinyl siding or corner boards. Veteran California framer Don Dunkley likes to use the two-stud configuration for interior corners but wants beefier corners on exterior walls that will hold the strapping and bolts necessary to comply with earthquake regulations. The two-stud corner works equally well in 2x4 and 2x6 walls.

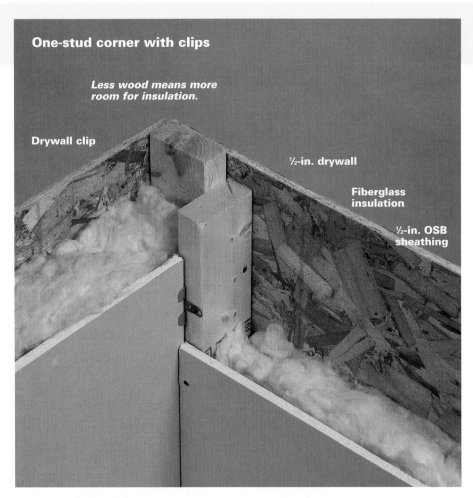

One-stud corner with clips

Less wood means more room for insulation.

Drywall clip

½-in. drywall

Fiberglass insulation

½-in. OSB sheathing

Built fast, one-stud corners adapt to newer methods

A variation on the two-stud corner, the one-stud corner is a direct result of the switch from drywall nails to drywall screws. The hammerproof 2x nailers in corners can be replaced with a thinner nailer (photo bottom left) or with drywall clips (photos top, bottom right) because screws need less backing support than nails. By nailing a 1x3 or a 3-in. wide strip of plywood to the inside corner, you eliminate an additional stud, thereby saving wood and creating room for insulation. Drywall clips (sidebar, p. 133) eliminate the need for wood nailers and further increase the available wall-cavity space for insulation. Instead of clips, Bill Eich of Spirit Lake, Iowa, has long strips of light-gauge steel bent lengthwise at right angles (resembling drywall corner bead), measuring approximately 1½ in. per side. His carpenters then screw the strips to the appropriate studs, where the strips serve as cheap, efficient drywall backers.

One-stud corner with nailer

1x3 or plywood strip

One-stud corner with light-gauge steel backer

1½-in. by 1½-in. light-gauge steel is screwed or nailed to stud.

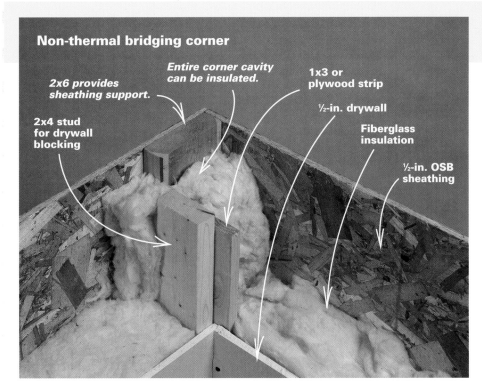

Non-thermal bridging corner

2x6 provides sheathing support.

2x4 stud for drywall blocking

Entire corner cavity can be insulated.

1x3 or plywood strip

½-in. drywall

Fiberglass insulation

½-in. OSB sheathing

A hybrid corner that eliminates thermal bridging

Contributing editor Mike Guertin has long been concerned with exterior corners, which can be notorious cold cavities. Although thermal bridging is a concern along the entire wall, framing alternatives such as double-stud walls are not cost-effective. Corners, on the other hand, carry less load than a typical stud and are structurally more flexible. Guertin is experimenting with a 2x6 corner (photo left) that can be completely insulated. He uses a 2x6 on the exterior corner and nails a 2x4 and a 1x3, or a strip of plywood, together to form the interior corner. He can now insulate throughout the corner cavity, eliminating thermal bridging by breaking contact between the outer sheathing and the drywall. Guertin says that the corner still provides good support for top plates and for nailing. (Although Guertin's building inspector approved the corner, check with your local inspector before trying it.)

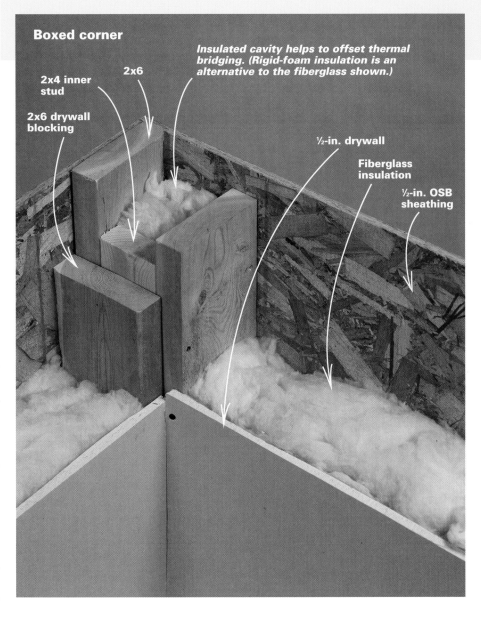

Boxed corner

2x4 inner stud

2x6

2x6 drywall blocking

Insulated cavity helps to offset thermal bridging. (Rigid-foam insulation is an alternative to the fiberglass shown.)

½-in. drywall

Fiberglass insulation

½-in. OSB sheathing

Boxed corner has mass and can be insulated

Here's another corner that is fairly popular (photo left). Typically used in a 2x6 wall, the outer studs used are 2x6s, while the inner stud can be a 2x4. John Carroll, a builder from North Carolina, uses a variation on this theme when nailing wider corner boards but uses another 2x6 in place of the 2x4. In both examples, framers should have insulation ready on site because the interior of the box must be insulated before the sheathing is nailed to offset the potential of thermal bridging. Builders who require more nailing for corner boards will sometimes put an additional stud on the outside of the box after insulating its interior (photo below).

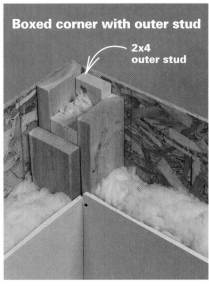

Boxed corner with outer stud

2x4 outer stud

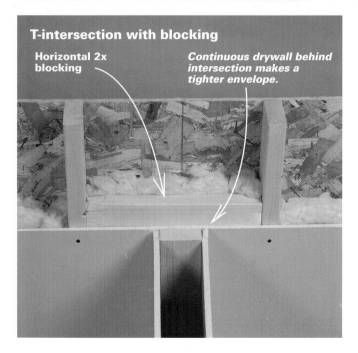

Three-stud intersection

½-in. OSB sheathing

Box made of 2x4s or 2x6s

On exterior walls, insulate box before sheathing.

Fiberglass insulation

½-in. drywall

T-intersections must provide good drywall-nailing support

A traditional method of building T-intersections is nailing the intersecting wall to a box, sometimes called a channel, built into the adjacent wall (photo left). To lessen thermal loss (if the adjacent wall is an exterior wall), this box must also be insulated before the exterior is sheathed. Blocks nailed horizontally between studs (photo top right) in the adjacent wall are a second method of forming the corner and a good place to use scrap lumber; the intersecting wall is then nailed to the blocks. Some builders prefer a faster third method (photo bottom left) and nail a wider (a 2x6 on a 2x4 wall, for instance) piece of stock on the flat that provides nailing for both the intersecting wall and the drywall in one shot. When the framing, insulation and drywall crews are working in close synch, it's also possible to drywall the entire first wall and then nail the new wall through the gypsum into the nailer. This method creates fewer breaks in the vapor barrier. Drywall clips do the work in the fourth method (photo bottom right).

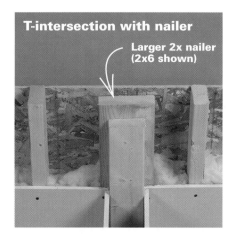

T-intersection with nailer

Larger 2x nailer (2x6 shown)

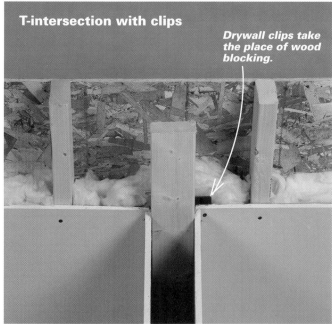

T-intersection with blocking

Horizontal 2x blocking

Continuous drywall behind intersection makes a tighter envelope.

T-intersection with clips

Drywall clips take the place of wood blocking.

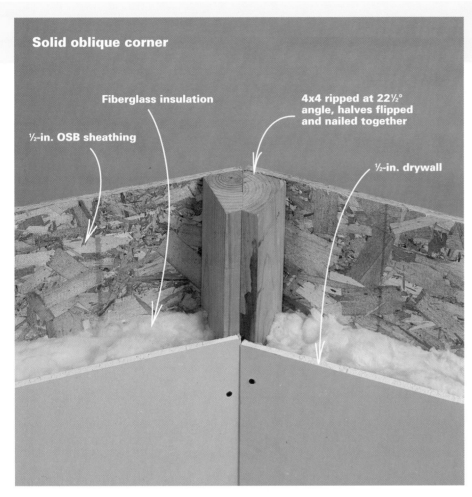

Solid oblique corner

½-in. OSB sheathing

Fiberglass insulation

4x4 ripped at 22½° angle, halves flipped and nailed together

½-in. drywall

Oblique corners can be a challenge to frame quickly

Corners greater than 90° present problems of their own. In the conventional method, studs aligned with their respective plates meet in a splayed fashion and are sometimes difficult to nail together into a straight corner. (Crooked corners can give drywall contractors fits.) Contributing editor Scott McBride uses a sturdier method (photo left) that appeared in the "Tip & Techniques" column of *Fine Homebuilding* (*FHB* #26). With a table saw or a circular saw, he rips a 4x4 in half at 22½°, flips one half around and nails the two halves together. If there are a fair number of these corners on a job, Don Dunkley will rip 2x4s at a 45° angle and nail the long blocks into the widest part of the cavity (photo below). This not only makes a larger nailing surface but also allows for insulation.

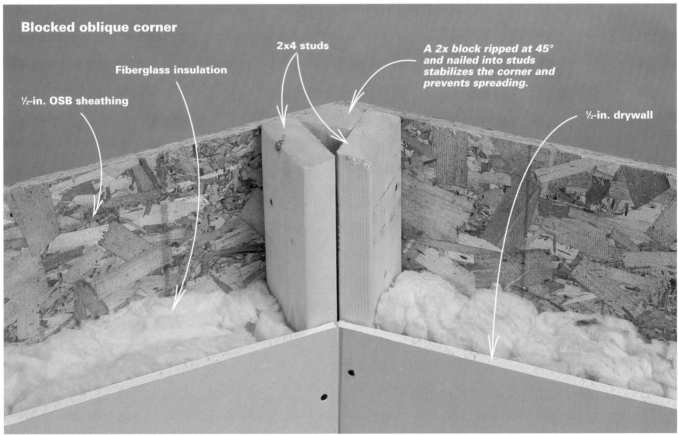

Blocked oblique corner

½-in. OSB sheathing

Fiberglass insulation

2x4 studs

A 2x block ripped at 45° and nailed into studs stabilizes the corner and prevents spreading.

½-in. drywall

Building an Efficient Fireplace

A variation of the Rumford design reflects heat into the room rather than letting it disappear up the chimney

by Richard T. Kreh Sr.

A couple of years back, my good friend Nick called to tell me that he'd bought a 40-acre tract of land in southern Pennsylvania. I wondered how this purchase would involve me, and the answer came quickly. Nick planned to build a cabin overlooking the valley, and the centerpiece of the cabin would be a large stone fire-

place. Having built hundreds of fireplaces as a mason, I was the right friend to call.

Finding stone wasn't a problem. The mountainside lot had many old stone fences and loose rock strewn everywhere. Nick, however, wanted this stone fireplace not just for aesthetic reasons but also for heating the cabin. After dis-

cussing the options, we decided that our best bet was a Rumford-style fireplace, efficient due to its wide, high, shallow firebox (photo above).

Rumford fireplaces radiate heat—Count Rumford was a talented inventor who fled the colonies to England in 1776 because he was a

Photo this page: Roe A. Osborn

Proportions for an efficient fireplace

- *The width of the firebox at the back and the depth of the firebox should be equal.*
- *The height and width of the vertical portion of the firebox back should be the same.*
- *The area of the fireplace opening (height times width) cannot exceed 10 times the flue opening.*
- *The height and width of the fireplace opening should be two to three times the depth of the firebox.*
- *The height of the opening should not be greater than the width.*
- *The throat should be 3 in. to 4 in. wide.*
- *The centerline of the throat should line up with the centerline of the hearth.*
- *The smoke shelf should be at least 4 in. wide.*
- *The height of the throat should be at least 12 in.*
- *A flat plate damper should be used at the throat.*

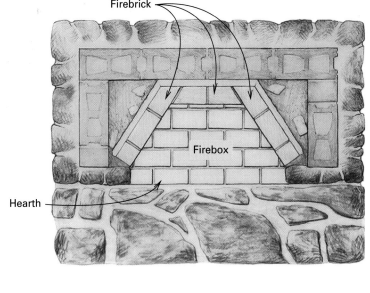

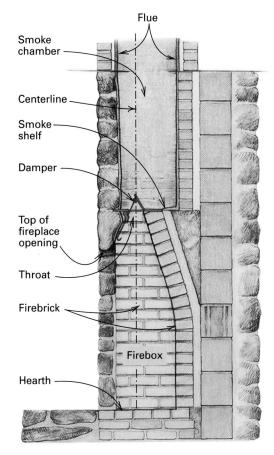

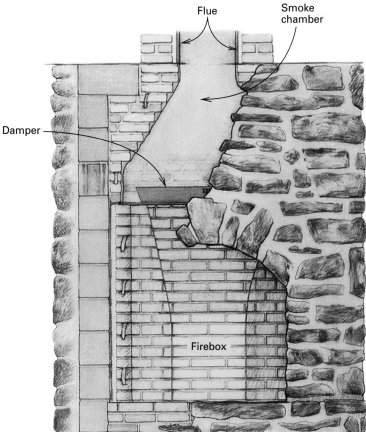

Drawing and formula adapted from Brick Institute of America

Loyalist and supported the monarchy. Rumford was fascinated with fireplaces and chimneys. He experimented with the principles of fireplace design and construction, and he wrote a number of essays and opinions on the faulty fireplaces he observed. He built a reputation in England by repairing hundreds of faulty fireplaces.

In almost every case, Rumford discovered that the throat area (drawing above) of the fireplace was too large; throats tended to be large because there had to be enough room for a chimney sweep to climb through to clean the chimney. He discovered that a large throat made a fireplace burn less efficiently.

In his study of fireplaces, Rumford also found that the most beneficial heat emitted from a fireplace is radiant heat, which is heat reflected into the room from the back and sides of the firebox. To maximize the effect of radiant heat, Rumford designed a fireplace with a tall, wide opening and a shallow firebox with widely an-

A concrete-block shell is laid up first. The first masonry that is installed is rough concrete block. The stone veneer will be attached to the outside of the shell while the firebox is built inside.

Angled firebox walls are sketched on the hearth. One key to a successful Rumford fireplace is the angled sidewalls of the firebox, which are laid out here on top of the firebrick hearth.

gled sides to radiate as much heat as possible. Rumford also built his fireboxes out of a smooth, fireproof material, such as brick, to promote the smooth flow of air and to reflect heat instead of stone, which has a rough surface that inhibits airflow and absorbs heat.

Getting the proportions right—As I began the plans for Nick's fireplace, I turned to Vrest Orton's book *The Forgotten Art of Building a Good Fireplace* (Yankee Inc., Dublin, N. H. 03444; 1974), one of the best books on Rumford-fireplace design and construction. In this book Orton outlines the details and proportions needed to build a successful Rumford fireplace.

I should point out that the design Orton describes in his book is actually an adaptation of Count Rumford's original historic design. Rumford's original design featured a firebox with a straight back and rounded breastwork, or front inside surface of the throat. Orton's variation, which is accepted by the Brick Institute of America in its publication *Technical Note on Brick Construction, 19C revised* (Brick Institute of America, 1988; 703-620-0010; $1) as the recommended Rumford design, incorporates a slight angle to the top half of the firebox back to increase the amount of radiated heat. Precast-masonry parts are available that form the rounded breast, but I decided against this option because I wished to build the fireplace of masonry.

According to Orton's formula (drawing p. 139), the width and height of the firebox opening can be up to three times the depth of the firebox. The back wall of the firebox should rise vertically from the hearth 16 in. to 20 in. and then slant or roll forward gently and evenly to form the smoke shelf at the top of the firebox. The damper should be approximately 4 in. deep with a smoke-shelf depth of at least 4 in., and the centerline of the throat and damper should align vertically with the center of the hearth. An-

The back wall of the firebox tips forward. To enhance the radiant properties of the fireplace, the top of the back wall of the firebox is rolled forward slightly.

other key factor in the optimum performance of the fireplace is making sure the throat and smoke-chamber areas are uniformly smooth and corbeled evenly with brick up to the point where the flue lining is set in place.

Roughing in the fireplace—Nick had seen to it that the proper footing and foundation had been built, so the first step for me was laying out the fireplace on the floor using the rough measurements in the Brick Institute's tables. We based our layout on a fireplace with a finished opening 40 in. wide by 40 in. high. Because the cabin was not airtight, we decided not to incorporate an outside air source into the fireplace.

The unfinished concrete-block shell went in first, allowing a 7-in. setback from the outside perimeter for the future stone veneer (photo top left, facing page). I embedded corrugated-metal wall ties in the mortar joints between every other course of block around the outside perimeter of the chimney to tie stone to block work.

I used the same mortar mix for all phases of the fireplace construction, including the block work, the firebrick and the stonework (some masons prefer a special refractory mortar for firebrick). The mortar mix I used is type N, consisting of 1 part type-1 portland cement, 1 part mason's hydrated lime and 6 parts sand with clean water added to get the desired stiffness.

The lime in the mix increases the mortar's ability to bond to the stone. At the same time, the lime decreases shrinkage of the mortar while imparting a nice light-gray color to the cured mortar. I recommend using only washed sand from your local building-supply dealer because of its uniform texture and because it gives the mortar a uniform color when it cures.

The firebox is laid out on top of the hearth—Nick wanted the hearth raised 8 in. above the finished floor. So I attached a guideline to the ends of the rough block at that height and laid two courses of rough brick and the outside course of hearth firebrick up to the line.

I put the head joint between the two middle firebricks on the centerline of the fireplace for a reference point and laid the rest of the hearth bricks front to back with mortar joints about ⅜ in. wide. I tooled the mortar joints with a convex half-round jointer. With the hearth floor complete, I penciled in lines for the first course of firebox bricks (photo top right, facing page).

If the finished opening of the fireplace was to be 40 in. wide, the firebox needed to be 20 in. deep, and the back of the fireplace had to be 20 in. wide. Because of the unevenness of the stone veneer, I adjusted the overall depth to 22 in. I laid the firebrick flat in what is known as the stretcher position, which made the firebox wall 4½ in. thick. This arrangement not only

makes for a stronger firebox but also keeps the firebrick from burning out prematurely. Laying the firebrick in stretcher courses also provides a greater brick mass, which acts as a heat sink to radiate more heat into the room. Stretcher courses do use more firebrick than if they'd been laid on edge, but the long-term benefits far outweigh any difference in price.

Building the firebox—The bricks for the firebox were laid up in a half-lap pattern, called running bond, to a height of 20 in. as prescribed in the table. At that point I began rolling or pitching the bricks for the back of the firebox toward the front (bottom photo, facing page) so that the top of the firebox would end up at the proper depth for the damper to be set at the throat.

To figure out how much to roll each course, I first figured out how far forward the wall would have to come to accommodate the damper. That distance worked out to be 8½ in., which I divided by 11, the number of courses needed to arrive at the height of the damper. Each course

Stringlines establish the corners of the stonework. Nylon strings stretched plumb from the ceiling to metal pins near the floor work as guidelines to keep the corners straight as the stone is laid.

therefore had to be rolled forward ¾ in. Rolling the firebrick forward caused each course in the back to be slightly lower than the side courses, so I had to add a layer of split firebrick to bring the bricks in the back even with the sides. Every few courses I embedded metal wall ties in the mortar joints on the front edges of the firebox to tie into the stonework later.

I parged the backside of the firebox brick with mortar to seal holes, but the angled sides of the firebox left large triangular voids between the back of the firebox and the rough block. I filled these voids with pieces of rough brick embedded in mortar, leaving a space of at least ½ in., or the thickness of a finger, between the firebrick and the rough brick to allow for expansion from heat. I loosely inserted a thin layer of fiberglass insulation behind the parged firebrick as I built it up to keep the mortar droppings out of the expansion space. Every couple of courses in height, I laid some metal wall ties in the mortar joints under the firebrick and into the rough-brick backing to tie the two together.

A custom damper is made from spare parts—I wanted to use a flat metal damper for Nick's fireplace, and none of the masonry-supply places in my area had what I needed. I wanted the damper to be made of plate steel with a lid that opened 90°. The lid of the damper had to lie on a flange that sat on the top edge of the firebox. To get the exact damper I wanted, I turned to a local metal shop.

I made a cardboard template the exact size of the damper so that the machinist could match the sides of the damper to the angled sides of the firebox. He made the damper flange and lid out of ½-in. plate steel to help prevent future warping from heat. I wanted the damper lid to open from the back instead of pivoting on a hinge or rod in the center of the lid; also, the damper lid had to be removable in case it should require service. The machinist's simple solution was welding a stop strip on the rear flange that the lid could ride against as it opens.

Next we needed something to hold the lid open while the fire was burning. The machinist offered to make something from scratch, but I ended up digging through a box of spare damper parts at the masonry-supply place instead. Sure enough, I found the notched arm assembly, and the supplier wouldn't even take any money for the parts. The machinist welded the parts in place, and we were in business. The damper cost me a total of $70. Back at the cabin, I centered the damper assembly on top of the firebox and installed it in a bed of mortar.

Corbeled brickwork supports the flue—With the damper in place, I was ready to start building the smoke chamber. I allowed for an

Fine-tuning the stone for a perfect fit. Most stonework requires some chiseling to get the stones to fit correctly. Here the author uses a standard mason's chisel to dress the face of a stone.

ample smoke shelf in back of the damper and built the back wall of the smoke chamber up to the required 29-in. height where the first flue lining would be set in place. For this wall as well as for the rest of the smoke chamber, I used hard, rough brick. I inserted some metal wall ties in the back wall at regular intervals to tie in the corbeled sidewalls.

Next I plumbed up with the level and marked the vertical centerline of the firebox on the top of the back wall of the smoke chamber. I knew from the table that I'd need a 16-in. by 16-in. flue lining for this fireplace, so I measured over 8 in. on each side of the center mark. From these points I struck a chalkline down to the top of the damper where the corbeling would begin. These lines gave me an even slope to follow as I built the sidewalls of the smoke chamber.

Before I started the sidewalls, I laid some fiberglass insulation around the ends and edges of the damper so that the damper had room to expand without cracking the masonry. Then I corbeled the brickwork up to the level of the flue lining. Corbeling is done by laying the bricks lengthwise toward the middle of the smoke chamber and letting every course project slightly beyond the course below following the snapped line. I made sure that all my mortar joints were solid and well filled. After the corbeling was finished, I rubbed down the mortar

joints with a piece of burlap to smooth the mortar and to fill any holes.

Next, the flue lining was set on the corbeled edge for support. For this chimney, I had to cut the bottom of the first lining at a slight angle so that it passed through the roof a safe distance (2 in. minimum) from any combustible material such as the roof and ceiling framing. The 2-in. space was filled with fiberglass insulation. I built the chimney from that point up through the roof the same as a conventional chimney.

Starting the stonework—With the chimney and the guts of the fireplace finished, I was ready to install the stone veneer. First I established the corners for the stone veneer by dropping plumb lines from the ceiling out 7 in. from the rough masonry. I laid a couple of courses of stone in mortar for each corner, then attached a tight plumb vertical line from the ceiling joists to a metal peg inserted between the stone courses (photo p. 141). These more permanent lines then became corner guides for the stone.

Mortar tends to set more slowly in stonework than in brickwork because the bricks absorb more moisture from the mortar. So I made the mortar for the stonework more stiff than the mortar for brickwork. I attached horizontal guidelines between the corners as I worked my way around the fireplace. The trick is to allow

enough time for the mortar to set up so that the stones don't sag, move around or push out. The squarest stones were used for the fireplace jambs and for the corners, and I intentionally combined different sizes to give the stonework a random appearance.

After the mortar had set slightly, I tooled the exposed joints by rubbing them with a short length of broom handle with a rounded end. I filled in holes or voids with mortar using a flat pointing trowel or flat jointing tool, pressing the mortar in firmly and then going back over the joint with the broom handle. After the mortar had set enough so that it would not smear, I brushed the joints gently with a soft-bristle brush for the final finish.

No matter how much stone you have to pick from, you'll always have to do some cutting or trimming. The stone for Nick's fireplace was no exception. I do all my stone-cutting on a heavy bench that I made out of framing lumber.

Minor trimming can be done with a brick hammer, but I use stone-cutting chisels and a hammer for most of my cutting (photo left). Stone chisels come in a variety of shapes and sizes, each with a specific purpose. Discussing all of them and how they are used is beyond the scope of this article. A basic set of stone chisels would include a pitching chisel that has a flat, slightly beveled blade used for squaring and dressing stone; a standard all-purpose mason's chisel with a wide bevel-edged blade; a toothed chisel for roughing out flat areas of stone; and a pointed chisel for removing lumps of stone too small to be taken off with larger chisels. Stone chisels are available at most masonry-tool suppliers. Wear safety glasses with side shields or goggles while chiseling stone; a glove to protect the hand holding the chisel is also a good idea.

Stone arch is built on top of a plywood form—Nick wanted an arched fireplace opening instead of the standard square opening. The basic rule of thumb for designing a segmental self-supporting arch is to make the rise of the arch no less than one-eighth of the span. At 40 in. wide the rise for this arch had to be at least 5 in. at the center point of the arch. However, Nick wanted a more pronounced curve, so we increased the rise to 12 in.

With a projected finished height of 40 in. at the center of the arch, the height of the stone jambs on each side where the arch springs from worked out to 28 in. (40 in. minus 12 in.). The next step for us was building a form that would support the arch while it was being built (photo facing page).

To avoid a lot of complicated math, I began by laying out the 40-in. square finished opening of the fireplace on a half-sheet of ½-in. plywood. Next, I measured up 28 in. on each side and drove small nails at those points. Then I marked the center point of the top of the opening, which was the highest point of the arch. Then I turned the plywood on edge with the top of the arch facing down. I tied a length of nylon mason's line to one nail and drooped the free end over the other nail. I let the line sag until the middle of the line reached the center point or the top of the arch. With a sharp pencil, I followed the line without touching it to scribe the arch.

Next I cut out the arch with a saber saw and used it as a template to cut a second arch for the other side of the form. To make the finished form close to the same width as the 7-in. stone veneer, I made supporting legs out of 2x8s along with a spreader piece that fit between the springing points. I then nailed the plywood arches to the front and back of the form and cut a series of 2x4 blocks to fill in along the curve of the arch.

I cut the legs a little short and inserted wedges under each leg to get the top of the curved form exactly level and where I wanted it. I also cut a spreader that fit between the legs near the hearth to keep the form from moving. The wedges would come in handy when the arch and the front wall over it had hardened and cured. At that point I simply had to knock out the wedges, and the form would drop free from the stone arch.

With the form in place, I selected stone that fit the curvature as closely as possible. I laid the stones across the arch with a wedge-shaped keystone in the center. No mortar was used under the stones because this was to be the finished edge. I used small wooden wedges to keep the stones in place while the mortar set up. Because the back of the arch stones formed the breast of the fireplace, I used stones that tapered toward the bottom. When the fireplace arch was finished, I parged the inside face to smooth the breast for better fireplace performance and to seal holes.

Proof is in the performance—I continued the stonework up to mantel height, about 60 in. from the finished floor. I'd picked out three fairly square bracket stones to support the mantel and now placed one on each end and one in the center using a line stretched tightly between the corners to keep them level.

I let the stones project out about 5 in. to support the 4-in. thick pine mantel that Nick had found at a local sawmill. We held the bracket stones in place with 2x4 props until the mortar had set up. I then could complete the stonework up to the ceiling on all four walls. We decided to install a wood mantel on the back wall of the fireplace at the same height as the mantel in front. Even without a fireplace opening on that side, the back mantel added a nice decorative touch to the massive, solid stonework.

Arch is built on a plywood form. A form made of plywood and framing lumber supports the stone arch until all the stonework is finished and the mortar has cured. When the form is removed, the arch will be self-supporting. A string stretched between the corners acts as a reference to keep the stone straight across the face of the fireplace.

Finally, we used large, flat stones to build a stone hearth that extends 20 in. out into the room. All the joints on top of the hearth were tooled flat so that they would not collect dirt.

I had protected the hearth with cardboard while the fireplace was being built and had been careful while applying the mortar, so the only cleaning that I had to do was removing a few particles or splatters of mortar with a stiff brush and a rounded-off piece of wood. In a couple of places, I needed a chisel to take off some stubborn lumps of mortar.

Masons sometimes wash stonework with a weak solution of muriatic acid to remove mortar stains and discoloration, but because we'd been careful, we skipped this wet, messy process. After letting the fireplace cure for two weeks, we built a test fire in it, starting with a small fire and adding wood until it was built up to a moderate size. A test fire allows everything to heat up evenly and gradually. The fireplace drew well, and no smoke was allowed into the room.

Nick has been using the fireplace ever since as the main heat source for the cabin, and one chilly raw day last December, I went with him to the cabin. We built a fire, and within minutes the cabin was warm enough for us to remove our jackets. A half-hour later, the wall of the cabin opposite the fireplace was warm to the touch from the heat radiating into the room.   □

Richard T. Kreh Sr. is a master mason, a consultant and the author of Delmar Publishers' Masonry Series, Masonry Skills. *He is also a frequent contributor to* Fine Homebuilding's *"Q&A" column. Photos by the author, except where noted.*

A Vermont farmhouse with a twist. This 2,600-sq. ft. house in southern Vermont looks a lot like its more traditional neighbors. But truss walls that are packed with 1 ft. of insulation make this house far more energy efficient.

High Efficiency at Low Cost

Larsen trusses and thick insulation produce a warm house for $48 per sq. ft.

by Jim Young

I tried plenty of energy-efficient design schemes during the 1970s and 1980s and found that many of them focused on the collection, storage and distribution of heat energy. The designs, especially solar strategies, seemed dependent on machinery—even when the energy source was sunlight. Yet the heat was often allowed to migrate right back out of the house through air leaks and underinsulated walls. It became obvious to me that a house with high R-values and tight construction wouldn't need as much heat because it wouldn't lose much. It made sense to build a house that could rely on internal heat gains—the heat derived from people, lighting and appliances. If built correctly, such a house would remain comfortable with the simplest of heating systems.

I got to use this approach in 1990 when Tom and Ellen sought out my design company for help in building a house. The couple owned three acres on a small mountain in southern Vermont. As beautiful as it was, the location would take the full brunt of winter winds, and the owners could count on 8,000 heating degree days a year. They wanted a warm, comfortable and environmentally safe house that would not require them to monitor complicated mechanical systems. Tom and Ellen wanted the house to fit in their neighborhood. To me, that meant a typically unadorned Vermont farmhouse (photo facing page). And finally, they wanted to get their house built without a mortgage, which translated into a tight construction budget.

The design approach to their objectives was simplicity: thick insulation and tight construction that retained heat rather than strategies to gain heat. The 2,600-sq. ft. house that resulted met all of our expectations (floor plan p. 147). A big part of that success is due to a type of wall truss pioneered by Canadian builder John Larsen more than 10 years ago; using Larsen trusses allowed us to pack in 1 ft. of fiberglass insulation for an R-value of 50 in the walls.

Foundation and walls—Foundation walls are standard 8-in. poured concrete with a double layer of 2-in. extruded polystyrene glued to the outside. The insulation, with an R-value of 5 per in., totals R-20. Over the insulation went wire lath and stucco.

The guiding principle behind Tom and Ellen's house was to separate insulation and structural

Splayed window returns. **The 17-in. thick walls in the house required custom window returns. The sides of most of them are splayed at a 30° angle to bring more light inside. Photo taken at A on floor plan.**

functions by building two separate walls and letting each of them do one job properly.

We started with a 2x4 wall built with 16-in. o. c. stud framing. The 2x4s were sheathed with ½-in. plywood, just as they would be in a standard house. Subs could then wire, plumb and duct as needed. There is no insulation or vapor barrier in this wall; those go elsewhere.

Adding a vapor retarder—Once the 2x4 walls were in place, an air-vapor retarder was added to the outside of the building, over the plywood, just like housewrap (or air barrier). At the top of each wall, 18 in. of excess vapor retarder was stapled temporarily to the third-floor decking; it would later be joined to the vapor retarder coming down the underside of the rafters. By applying the vapor retarder in this fashion, we covered all the intersections of floors and walls in one simple step and avoided the gaps and the leaks common in houses where a vapor retarder is applied room by room, floor by floor. Because the vapor barrier in this house was applied so effec-

tively, I decided housewrap—to prevent bulk movement of air through walls— would have been superfluous. We skipped it.

Trusses for roof and walls—To protect the house from the weather, we tackled the roof as soon as the 2x4 walls and floors were complete. Because the third-floor was designed as a warm space, I used 16-in. deep roof trusses so that we could pack in plenty of insulation. The trusses, which were made by a company in the area, were set 2 ft. o. c. and allowed 14½ in. of insulation with a 1½ in. vent channel above.

Venting on this roof included a typical ridge and soffit system, but we had some difficulty finding a product that would maintain the 1½-in. space beneath the roof sheathing when the fiberglass insulation was blown in. Our insulation contractor warned us against molded foam vent channels because they might deform under pressure, possibly blocking some of the airflow from eaves to ridge. Instead, we nailed 1½-in. wide strapping to the top inside edges of the roof trusses, then built our channels from ¼-in. interior-grade plywood that wouldn't buckle when insulation was blown in.

With the roof in place, it was time to apply the wall trusses to the outside of the building (for more on how we built the trusses, see sidebar on p. 146). The Larsen trusses made up the second of our two outside walls; the second wall is devoted to insulating the house. Trusses were set 2 ft. o. c. and nailed into the rim joists or headers over the windows to anchor them to the wall (top photo, p. 146). At the top of each truss, the plywood web was extended 1½ in., leaving a flange that could be nailed directly to the horizontal soffit member of each roof truss. At the bottom of the wall, the rim joist had been held back 2 in. from the edge of the mudsill. After the CDX sheathing was applied, a 1½-in. ledge remained. We took 1-ft. wide sections of treated plywood and nailed them to the ledge, then set the wall trusses down on the plywood. When the plywood was nailed into the truss chords from below, the wall cavities were sealed (bottom photo, p. 146).

Making window openings—Window openings required custom detailing unique to thick-wall construction. Walls measure 17 in. from drywall to outer sheathing; to avoid a fortresslike feeling and to let in more light, the sides of the window

Trusses in place. This view shows the house after all the Larsen trusses have been nailed to the outside of the house and all window openings framed in 2x4s. The vapor retarder is between the trusses and the plywood sheathing; the red lines are taped seals between sheets of an 8-mil vapor-retarding polyethylene.

Sealing truss walls. Strips of pressure-treated plywood seal the bottom edge of the truss walls. The plywood is nailed to the outside edge of the mudsill, then nailed to the bottom of each truss. The red strip is construction tape that bonds the vapor barrier to the plywood.

Truss system makes thick insulation possible

The truss system I used on the house in southern Vermont was developed more than a decade ago by John Larsen, a builder in Alberta, Canada. The system is simply a series of lightweight racks that hold large amounts of insulation. They can be used in new construction or for retrofits (for more on the Larsen truss, see *FHB* #20, pp. 35-37).

Trusses can be made in almost any depth; we settled on 1 ft. because that allowed as much insulation as we needed. As Larsen originally built them, the trusses were made from 2x2 vertical chords that had been dadoed to accept 1-ft. squares of ⅜-in. plywood spaced every 2 ft. or 3 ft. I made the trusses a little differently: first, by using 2x3s straight from the lumberyard instead of ripping 2x2s from larger stock, and second by skipping the dadoes in the truss chords. The carpenters working on the Vermont house attached 1-ft. squares of ½-in. plywood to the sides of the 2x3s with waterproof yellow glue and ring-shank nails.

We set up a jig for the most common length we'd need—a truss that would stretch 19 ft. from the foundation to the top plate. With the jig holding two 2x3s exactly 1 ft. apart, we then marked for the plywood squares about every 3 ft. That distance could be adjusted depending on window openings. Some of the trusses ran between window openings and were shorter than 19 ft.; for those longer than 19 ft. (on the gable ends), we spliced together shorter sections with drywall screws and blocks of wood as the trusses were installed. As the trusses were assembled on the jig, they were labeled and then stockpiled for use.

Trusses were nailed on the walls through the 2x3 chords into headers and rim joists 2 ft. o. c. Along top and bottom outside edges, the trusses were joined to each other with short lengths of 2x3s toenailed at each end into the 2x3 chords. Outside corners were a little tricky. An L-shaped assembly made from 2x6s was attached to trusses on both sides with 2x3s to make a solid outside corner. To anchor each truss at the top of the wall, the uppermost plywood gusset was extended 1½ in. and nailed to the adjacent roof truss. At the bottom of each truss wall, sections of plywood were nailed in from the bottom to complete the box that would hold the blown-in fiberglass insulation.

The Larsen trusses are very strong. When nailed to the house, they became a sort of ladder or scaffolding that wouldn't budge, even when we jumped up and down on them. Although the original purpose of the truss system simply was to hold insulation, it became apparent that it adds a lot of rigidity to the building envelope, too.

The added net cost of using the Larsen truss system was $6,562, or just over 5% of total costs. Here's a breakdown of the added construction and insulation costs plus a summary of how much the system saved in comparison with typical wood-frame construction. —*J. Y.*

Extra costs

Additional insulation:	$2,770
Materials for trusses:	$2,130
Truss roof system:	$1,100
Extra foundation insulation:	$662
Additional labor on trusses:	$7,000
Extra detailing due to trusses:	$1,000
Miscellaneous:	$600
Total additional cost:	$15,262

Credits due to truss system

Saved costs of heating system:	$7,200
2x4 walls instead of 2x6:	$900
Saved labor on vapor barrier:	$100
Housewrap not needed:	$500
Total credits:	$8,700

Net difference: **$6,562.**

Photos this page: Steve Mindel

openings are splayed 30° (photo p. 145). This means the rough openings in the 2x4 wall are 14 in. wider than the rough openings in the outer wall. Window openings in outer walls were framed with 2x4s to carry the window units.

The wall framing, the windows and the vapor retarder were indeed fussy details that required a lot of forethought and a little practice, but in the end they went smoothly. The finished windows have a nice feel to them, and the wide windowsills leave plenty of room for plants and cats.

The windows are argon-filled, low-e double-hung units (Marvin Windows, Box 100, Warroad, Minn. 56763; 800-346-5044). Even though they are of high quality, they are still the weak link in the thermal envelope. Adding window insulation to the house probably could cut heat loss in half, but the owners have so far decided not to bother because their heating bills are so low already that there isn't much money to be saved.

Insulation and ventilation—My two insulation options were loose-fill cellulose or short-fiber fiberglass; each has an R-value of about 4 per in. Fiberglass was our choice because of the personality of the contractor and the price quote (see sidebar facing page).

We were ready to blow insulation into the house after the exterior walls had been sheathed (but not sided), the windows were installed, the roof venting complete and the vapor barrier installed. It also made sense to strap the ceiling in the attic to help contain the insulation during the blowing process. In addition, we sealed any wall penetrations.

The insulation installer crawled into the small attic cavity beneath the ridge and dropped a 3-in. hose down the slope of the rafters and into the walls as far as he could. As he pumped the fiberglass into each cavity, he slowly pulled up the hose. After that, he went outside and drilled some 1½-in. holes in the sheathing to pump in more insulation and to bring each bay up to the required density. We achieved R-50 in the walls, R-58 in the sloping part of the roof and R-60 in the cap. Siding was applied after that.

Mechanical ventilation—Fresh air for occupants is provided by a Van-EE air-to-air heat exchanger (Conservation Energy Systems, 2525 Wentz Ave., Saskatoon, Saskatchewan, Canada S7K 2K9; 306-242-3663) that provides fresh air to every room in the house and exhausts the bathrooms and the kitchen. During the winter about 60% of the heat in the exhaust air is transferred to incoming fresh air. Tom has adjusted the exchanger so that it's providing an estimated .25 air changes per hour, a level he and his family find comfortable. Because the house is so tight, fresh air must be piped directly to each appliance that needs it: the woodstove, a gas dryer and a gas water heater.

Energy performance—Including the third-floor, the house measures 2,600-sq. ft., and much of the heat is provided by people, lights and appliances. When additional heat is required during the winter, Tom and Ellen fire up a small cast-iron woodstove in the living room. But on sunny days, the woodstove isn't needed. And because there is so little heat loss through the walls, the heat generated inside the house has plenty of time to migrate through the building, so there is never more than a 5° variation in temperature from room to room. Inside temperatures during the winter are usually between 68° and 70°. The total annual heating bill is $225, enough to buy 2½ cords of wood.

Even if the house were left vacant for the winter, indoor temperatures would stay above freezing. Tom's insurance agent had a hard time believing that, so he was reluctant to write a policy for the house. But Tom and Ellen proved their point the winter before they moved in, when the house was closed to the weather but left unheated. In late fall, the inside temperature was measured at 40°, and that's about where it stayed for the winter, even when outdoor temperatures dropped to -20°. We know because we installed a temperature sensor that could be checked by telephone. And all winter, Tom and Ellen, the carpenters and I called to find out how cold it got inside. Now the owners, and their insurance agent, are sure that even without power, the house is virtually freeze-proof. □

Jim Young runs Young Design, a design company in Bath, Maine. Photos by Scott Gibson except where noted.

SPECS

Bedrooms: 3
Bathrooms: 2
Heating system: Woodstove
Size: 2,600 sq. ft.
Cost: $48 per sq. ft.
Completed: 1991
Location: West Dummerston, Vermont

First floor

A
Chimney

Double door
to future porch

Second floor

Dn
Entry

Insulated Larsen truss wall

Uninsulated 2x4 bearing wall

North

0 2 4 8 ft.

Floor-plan key

1 Master bedroom
2 Family room
3 Kitchen
4 Dining room
5 Mudroom
6 Pantry
7 Bedroom

Photos taken at lettered positions.

Thick walls with a simple floor plan. *A house designed for long Vermont winters uses double-wall construction and a simple floor plan. The 2x4 bearing wall contains plumbing, wiring and air ducts while the truss wall outside has room for 1 ft. of blown-in fiberglass insulation.*

A Solar House in a Cold Climate

A well-insulated, solar-heated house braves New England winters with only a domestic-water heater as backup

by Marc Rosenbaum

Late in 1991, I was approached by a New Hampshire couple who wanted to build a house that would push the envelope of energy-efficient construction. At the same time, they wanted a house that would be comfortable and durable, and that would cost no more than other custom homes in the area. And they didn't want the house to look as though it had just landed from outer space.

As a designer of superinsulated, passive-solar homes, I was grateful for the opportunity to see how close we could get in the cold New Hampshire climate to a home that uses no supplemental energy for heat and hot water. I also wanted to be vigilant about the energy that would be used used for lights, appliances and mechanical systems, and to incorporate as many other environmentally friendly features as possible, emphasizing resource-efficient construction materials and a healthy indoor environment.

My clients had chosen a site not far from the center of Hanover, N. H. The design program included three bedrooms, two baths and a library. One of the bedrooms would be used as a home office. An open-plan kitchen, dining/living room, full basement, two-car garage and entry porch rounded out the list of spaces. We were looking at 1,700 sq. ft. to 1,800 sq. ft. of living space. The clients contracted the house themselves and hired Jay Waldner of Weewaw Construction in Plainfield, N. H., as the builder.

Build tight, ventilate right—In the northern New England climate, the sun is a welcome but occasional visitor, so a solar home first must have a building envelope that reduces heat transfer to a minimum. The required elements are airtight construction, plenty of insulation and high R-value windows.

Assembling the solar collector. A roof-integrated solar collector, air-sealing and efficient, south-facing windows are among the components of a house designed to function as an energy-saving system.

Air-sealing was a top priority on this project. Penetrations were minimized and carefully sealed, typically with foamed-in-place urethane. Bottom plates were sealed to the subfloor with a flexible sealant made by Tremco (3735 Green Road, Cleveland, Ohio 44122; 216-292-5000), and the walls and ceiling were covered with polyethylene sheeting, carefully lapped and sealed. The sheeting doubled both as vapor retarder and air barrier. The ceiling poly was installed in unbroken sheets on the trusses before the interior partitions were framed. Neoprene roof boots were used to seal the plumbing stacks and the radon stacks where they penetrated the ceiling air barrier.

Blower-door test results projected an infiltration rate of only 0.03 air changes per hour, many times tighter than typical new construction. (For more on blower doors, see *FHB* #86, pp. 51-53.) With a house this tight, mechanical ventilation is called for, and steps should be taken to reduce the level of indoor pollutants. (More on this later.)

Nearly a foot of insulation—Computer modeling told us that we needed R-values of about 40 for the walls and rim joists and 60 for the roof (drawing facing page). We achieved these R-values by building two 2x4 walls 4½ in. apart and filling the resulting 11½-in. cavity with cellulose insulation. On the basement side of the rim joists, we installed 2 in. of extruded polystyrene foam followed by an R-30 fiberglass batt. The roof is mostly trussed and is insulated with R-60 blown-in cellulose. The concrete basement is finished with 2x4 stud walls insulated with R-11 fiberglass batts, and there is 1 in. of extruded polystyrene foam under the slab.

Insulating windows glazed on site—The third component of superinsulation is the glazing, and here we made significant improvements at little or no increase in cost. When low-emissivity (low-E) glass with inert-gas fill is used in wood or metal windows, the sash and frame can have lower R-values than the glass. For this house, we chose Owens Corning windows (Owens Corning, Fiberglas Tower, Toledo, Ohio 43659; 800-438-7465) in which the frame and sash are composed of hollow, rigid fiberglass filled with rigid-fiberglass insulation. This reduced heat loss through the nonglass portions of the windows. (Owens Corning no longer makes fiberglass insulated windows for new construction. Still available is an insulated replacement window that can be retrofitted with nailing fins.)

Owens Corning didn't offer windows with R-values as high as we wanted, so we asked them to sell us casement windows without glass that we could glaze on site. Into the Owens Corning frames we installed prefabricated, sealed glazing units made by Southwall Technologies (1029 Corporation Way, Palo Alto, Calif. 94303; 800-365-8794). Each glazing unit contains what Southwall calls Heat Mirror, a polyester film with a low-E coating that is suspended between two layers of glass. We specified a low-E glass called Energy Advantage by LOF (Libbey Owens Ford Co., 811 Madison Ave., P. O. Box 0799, Toledo, Ohio 43697; 419-247-3731) for the inside pane and clear glass for the outside pane; we had the interior spaces filled with krypton, a low-conductivity inert gas. The combination of these elements gave us excellent insulating value, about R-6.5, without cutting down too much on solar transmission. We wanted to allow sunlight to pass through the windows because it helps to heat the building. Three-quarters of the glass is on the south side of the house.

Structural framing supports the large panes—The five fixed-pane units on the front of the house (photo facing page) also use Heat Mirror. In this case there are two layers of polyester film between two layers of clear glass. Southwall calls this product Superglass, and it's aptly named: With the krypton-gas fill, it has an R-value of 9. These units, measuring 46 in. by 76 in., were set directly onto the wall framing.

We also used Superglass in an exterior door on the southwest corner of the house and in a Wasco skylight (Wasco Products Inc., 26 Pioneer Ave., Sanford, Maine 04073; 207-324-8060) over the stairs to the second floor. Site-glazing Superglass into insulated-steel doors can produce high R-value doors at a reasonable cost.

When the dollars settled at the end of the project, my clients paid little more, if any, for the windows and doors than they would have had they used premium-brand windows. And they got significantly better thermal performance. I estimate the superinsulation, including air sealing, insulation and glazing, added about $6,000 to the construction cost.

Sizing the solar system—Because of superinsulation, the house has a heating load of about 20,000 Btu per hour (approximately 6kw) at -25°F. The house could be heated by four good-size blow dryers, but neither of the occupants has big hair, so we needed another strategy. We analyzed heat and domestic hot-water (DHW) systems, looking at their costs in terms of installation, annual operation and replacement.

In the end, we chose a heating system that could be described as hot-water heated forced air (drawing p. 150). We considered radiant heat, but that would have required a higher water temperature, which in turn would have meant a larger collector and storage tank. Using computer-aided modeling, we determined that a solar roof collector of 360 sq. ft. was required with

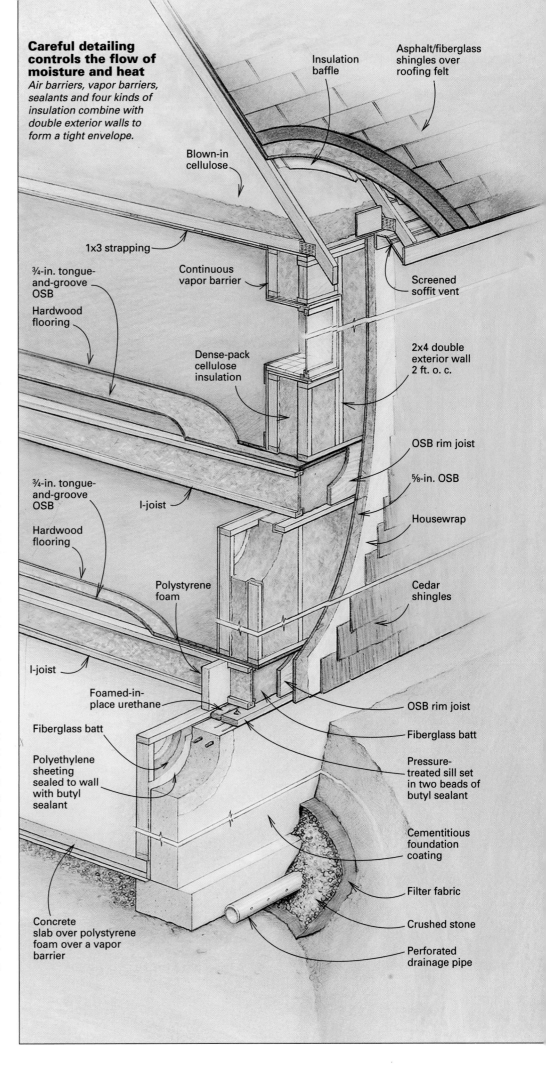

Careful detailing controls the flow of moisture and heat
Air barriers, vapor barriers, sealants and four kinds of insulation combine with double exterior walls to form a tight envelope.

Insulation baffle

Asphalt/fiberglass shingles over roofing felt

Blown-in cellulose

1x3 strapping

¾-in. tongue-and-groove OSB

Hardwood flooring

Continuous vapor barrier

Screened soffit vent

Dense-pack cellulose insulation

2x4 double exterior wall 2 ft. o. c.

OSB rim joist

⅝-in. OSB

Housewrap

¾-in. tongue-and-groove OSB

Hardwood flooring

I-joist

Polystyrene foam

Cedar shingles

I-joist

Foamed-in-place urethane

OSB rim joist

Fiberglass batt

Fiberglass batt

Polyethylene sheeting sealed to wall with butyl sealant

Pressure-treated sill set in two beads of butyl sealant

Cementitious foundation coating

Filter fabric

Concrete slab over polystyrene foam over a vapor barrier

Crushed stone

Perforated drainage pipe

1,200 gal. of water storage. We installed a site-built array of collectors that replace the roof surface instead of separate collectors over the roofing material (top photos, facing page). Flashing similar to that for skylights makes the transition from collector to roof. This approach saves cost, looks better and is more efficient. Roof-integrated collectors have less edge area to lose heat.

A copper collector with low-iron glass— The heat-absorbing portion of each collector panel consists of a 4-ft. by 10-ft. grid of copper tubing with a sheet of copper connecting the individual tubes. A coating that reduces radiant-heat loss is plated onto the surface of the copper (photo top right, facing page). The copper is enclosed by a single layer of solar glass that has a low iron content, allowing more of the sun's energy to pass through. The copper absorber and the glass both were supplied by American Energy Technologies Inc. (P. O. Box 1865, Green Cove Springs, Fla. 32043; 800-874-2190).

The system is the drain-back type: When there is no sunlight to warm the collectors, water drains from the roof into the storage tank in the basement (drawing below). When the sun returns, a sensor mounted on the absorber indicates that there is heat to collect, and a circulating pump is activated. Water from the bottom of the basement storage tank is pumped through the roof-mounted collector (photo center right, facing page), and it returns by gravity to the top of the tank. The house can be heated with 102°F water at an outdoor temperature of -25°F.

The 1,200-gal. storage tank (photo below) was made on site from sheets of copper soldered together, creating a highly corrosion-resistant tank that is not difficult to repair. The tank, about 6 ft. in dia. and 6 ft. high, cost less than $1,000, including labor and material. It's insulated from below with rigid foam and on the sides with two layers of R-19 fiberglass batts.

Water warmed on its way to the heater— Domestic hot water is stored in a separate 52-gal. water heater lined with polybutylene and insulated with 2 in. of foam. As hot water is drawn from the domestic tank, fresh makeup water passes through a heat-exchanging coil immersed in the top of the solar storage tank. To keep the temperature up in the DHW tank, water can be circulated through the storage-tank coil, which is made of 50 ft. of ¾-in. dia. copper tubing. The 5kw electric element in the domestic-water heater is the only source of heat outside the solar hot-water system.

There are both supply and return ducts in each bedroom to ensure good heat distribution and to keep pressure balanced. The ducts were sealed with latex mastic (RCD Corp., 2850 Dillard Road, Eustis, Fla. 32726; 800-854-7494) and insulated in the basement.

Air inside is clean and circulated—Several factors contribute to the house's healthy indoor environment. No gas or oil is burned inside; many of the paints and finishes are water-based; high-quality, natural finish materials are used throughout (photo bottom right, facing page), including granite, ceramic tile, linoleum and native hardwoods (instead of plastic laminate, synthetic carpet, vinyl and particleboard). A passive stack radon-mitigation system will make radon control easy if a radon fan is needed.

Finally, fresh air is provided mechanically, via the return ductwork, by a Van EE 1000 heat-recovery ventilator (Conservation Energy Systems, 2525 Wentz Ave., Saskatoon, Sask., Canada S7K 2K9; 800-667-3717). A heat-recovery ventilator, sometimes called an air-to-air heat exchanger, is a ventilation unit that includes a fresh-air fan, an exhaust-air fan and a heat-recovery core. The core is designed to allow the warm exhaust-air stream to transfer most of its heat to the incoming fresh-air supply while keep-

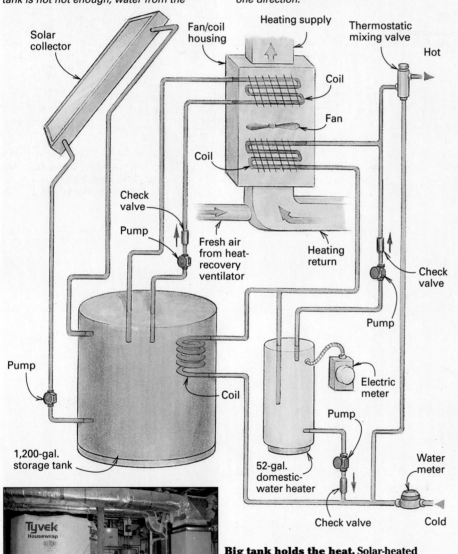

A simple solar system

Water is pumped to the solar collector on the roof from the 1,200-gal. storage tank in the basement. Heat from the water is transferred in the fan/coil housing to the air that heats the house. If the water in the solar storage tank is not hot enough, water from the domestic-water heater is pumped to the fan/coil housing. Fresh water can be prewarmed in a coil in the storage tank on its way to the domestic-water heater. The check valves limit the flow of water to only one direction.

Solar collector

Fan/coil housing

Heating supply

Thermostatic mixing valve

Hot

Coil

Fan

Coil

Check valve

Pump

Fresh air from heat-recovery ventilator

Heating return

Check valve

Pump

Pump

Electric meter

Pump

Coil

1,200-gal. storage tank

52-gal. domestic-water heater

Water meter

Check valve

Cold

Big tank holds the heat. Solar-heated water is stored in a site-built, 1,200-gal. tank in the basement. The tank, about 6 ft. in dia. and 6 ft. high, was made of sheets of copper soldered at the seams and is wrapped with insulation and an air barrier.

Solar collector built in place. The first step in building the roof-integrated solar collector was to install a frame or curb of 2x6s on edge.

Insulation reduces heat transfer to roof. Fiberglass batts were installed between the roof deck and the copper collector.

Sweated copper tubing links collector panels. The black coating plated on the surface of each collector reduces radiant-heat loss.

ing the streams separate to avoid contamination of the fresh air. Efficiency of residential units is typically between 60% and 80%.

Annual energy bill of $450—The house has been occupied for more than a year now. During the first year, the house used 1,197kwh for backup heat and domestic hot water, which is the equivalent of about 45 therm. of natural gas or about 35 gal. of fuel oil. Energy use for the year—lights, heat, etc.—was 4,255kwh, at a total cost of $450.

As I see it, the annual energy savings pay for the superinsulation upgrades. The solar/mechanical systems in the house, including heat, DHW and ventilation, cost about $15,000. The typical home in this part of the country would have an oil-fired, forced hot-water heating and DHW system, a woodstove and a two-flue masonry chimney and hearth, at a cost of about $14,000. With a heat-recovery ventilator, the price goes to $17,000. I don't think our mechanical system cost more than what is spent in this area. We just spent the money differently.

Room for improvement—Energy consumption could be further cut by a number of measures. More efficient motors in the solar-system pumps and the fan coil blower could likely save 250kwh to 300kwh annually. Light comes from a range of sources. Some are efficient, such as halogen and compact fluorescent bulbs, and some, chosen by the homeowners for aesthetic reasons, are less so. A superefficient refrigerator might save 300kwh to 400kwh. Another 150kwh to 200kwh might be cut by replacing the desktop computer in the home office with a laptop. In actuality, these projected savings would not be fully realized because some of the electricity used by lights and appliances helps to heat the house during the coldest months. □

Marc Rosenbaum, P. E., a designer of energy-efficient and environmentally sound housing, owns Energysmiths in Meriden, New Hampshire. Photos by Reese Hamilton, except where noted.

A house that draws energy from its environment. Water is pumped to the solar panels on the roof only when the collectors sense that there is heat to collect. This way, antifreeze is not needed to keep the water from freezing when the sun goes down and the temperature drops.

Safe, durable and attractive. The owners finished the inside of the house with materials that create little indoor pollution, such as hardwoods, granite and ceramic tile. The thickness of the highly insulated exterior walls can be seen at the window openings.

Cozy in Any Weather

Creating comfortable spaces in a home that faces New England's harsh winters and hot summers

by Linda Moody

The house that my husband and I built in rural Massachusetts (photo above) continues the evolution of New England architecture. It began with the first saltbox designs. The early settlers expanded the basic gabled-roof house design by adding more rooms at the back to make living indoors through the long, harsh winter easier. Completed in 1988, our house has typical New England features: wood clapboards, two-over-two windows and gables to maximize second-floor space. The colors we selected for the exterior come from nature: cedar stain on the wood

clapboards, bright hunter green for the trim. The house is practically camouflaged in spring, summer and fall in the screen of oak, black birch, maple and shagbark hickory trees.

Nature influenced the design of our house in other ways as well. Like the early settlers before us, we had to contend with the New England weather. I designed the house to be both warm in the winter and cool in the summer.

The garden room—We began by siting the house so that its long axis faces 15° off true south

(drawing p. 153). The saltbox shape works to buffer cold northwesterly winds. The roof comes down low on the northwest, from a 12-in-12 pitch to a 4-in-12 pitch, thus channeling prevailing winter winds up and over the house. We then located most of the windows on the south and east elevations; the north has only one window. We used mostly double-hung, single-pane windows with true divided lites and exterior glass storm panels set into the wood sash. The storm panels kick up the R-value to about that of insulating double-pane glass. Triple-pane glass, though it has a greater R-value, reduces solar gain, and we wanted to use the sun's energy to help heat our house.

The house's siting also dictated the layout of rooms, especially our garden room (top left photo, p. 154). Located on the southwest corner of the house, the garden room is a two-story passive-solar sunspace. It has about 90 sq. ft. of south-facing, double-pane casement windows (which, unlike double-hung windows, can be opened completely) and a stone floor. The floor, a combination of bluestone and crushed stone, has high thermal mass to help regulate indoor temperature: It slowly absorbs and stores the sun's heat and then disperses it at night. The floor consists of 6 in. of crushed stone over a layer of 2-in. rigid insulation that lies over several feet of clean gravel. The insulation prevents stored daytime heat from escaping downward.

Some of the plants sit on the stone floor in pots and in flats. Crushed stone, in addition to having great thermal mass, allows water to filter downward and evaporate, so these plants can be watered without worrying about floor damage or cleanup. And instead of wood baseboard, I laid up a course of colorful Great Wall tiles, which were hand made in China. Each tile looks like a watercolor painting. I used more of these tiles on a kitchen countertop.

The garden room is an oasis in all seasons, and it has given rise to a new family tradition. On New Year's Day, we gather there and plant seeds in flats for our gardens. This makes spring feel a little bit closer, and indeed with this room, plants and flowers always feel a part of the house.

On the second floor, the master bathroom (right photo, p. 154) features a whirlpool surrounded by double-hung exterior windows and three more casement windows that open to the garden room. The windows overlooking the garden room create the feeling of bathing in a quiet pool in the woods and allow the garden's solar heat to reach the upstairs bedroom through the bath.

Lighting enhances comfort—The house's design also takes advantage of natural light. The kitchen and the bedrooms face east to pick up the morning sunshine. The kitchen has solid-maple and Great Wall tile countertops, walnut and maple cabinets and a wood ceiling. With early morning sun shining in, the room is cheerful and homey.

Skylights in the kitchen and in the living room let sunlight into the house. In winter, with the leaves gone from the trees, the low sun shines through the doors of the west-facing porch and beams on our massive brick masonry woodstove, highlighting its rustic colors and giving a sense of warmth and comfort to the interior. The porch screens feature my own touch-fastener-type system, making them easy to remove for maximum winter light.

My intention with the house's design was to have an open floor plan but still allow for intimate spaces to make it cozy. So the floor plan includes several bump-outs, giving us special places, like the piano nook in the living room and the reading alcove off the dining room. The

SPECS

Bedrooms: 2
Bathrooms: 2
Heating system: Masonry woodstove
Size: 2,500 sq. ft.
Cost: $66 per sq. ft.
Completed: 1988
Location: Pepperell, Massachusetts

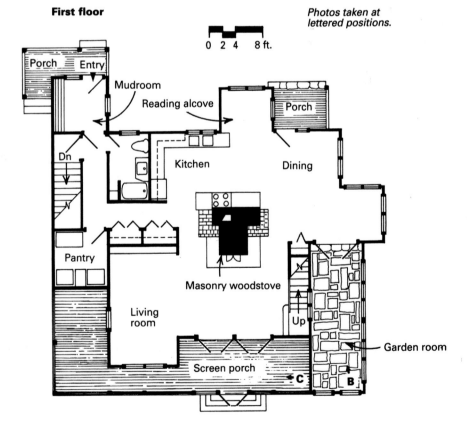

Second floor

Balcony

Master Bedroom

Bedroom

E

A

Loggia

Dn

Study

D

(Turret above)

◀ North

First floor

Photos taken at lettered positions.

0 2 4 8 ft.

Porch Entry

Mudroom

Reading alcove

Porch

Dn

Kitchen

Dining

Pantry

Masonry woodstove

Up

Living room

Garden room

Screen porch

C

B

An irregular footprint. *The house was sited to collect sunshine in the garden room; elsewhere, windows in bump-outs and a screen porch on the west elevation create a variety of lighting experiences.*

Walled-in pond. On the second floor, over-looking both the garden room and the sylvan exterior, this bathroom contains a whirlpool. Photo taken at A on floor plan.

With rocks and stones and trees. Inside the garden room, bluestone and crushed stone store the sun's heat and radiate it at night. This heat enters the second-floor bathroom through casement windows at the top of the far wall. Photo taken at B on floor plan.

Like the wind touching the sail. On the west elevation, the screen porch catches cool summer breezes coming off the Nashua River. The porch also shades this side of the house to control summertime overheating. The double French doors open into the living room. Photo taken at C on floor plan.

A variety of spaces. The living room, viewed from the loggia, is designed to be an entertaining area that is open to the kitchen and, through two sets of French doors, to the screen porch. Its expansive feeling contrasts with the intimacy of the piano nook at the far end of the living room. Photo taken at D on floor plan.

Bedrooms face east. Besides letting morning sunshine into the master bedroom, this triple French door unit offers a panoramic view that, combined with the cathedral ceiling, enhances the spaciousness of the master bedroom. Photo taken at E on floor plan.

reading alcove is surrounded by windows; it's a friendly, sun-filled spot to sit and read during the long, cold winter.

Varying ceiling heights and planes also makes each space different and lends intimacy to the open floor plan. The kitchen has a low ceiling of T&G, V-grooved 1x6 pine. This room is open to the living room (photo p. 155), which has a pine cathedral ceiling with operable skylights that help keep the entire house cool. An upstairs balcony/hallway, or loggia, overlooks the living room.

The master bedroom (photo above) has a cathedral ceiling with a skylight. A triple French-door unit topped with a transom and a half-round window leads outside to a balcony overlooking the gardens. Being able to see the outdoors upon entering makes this room feel larger. This is a trick used in condominiums that have limited square footage.

Holding onto the heat—You enter the house through a mudroom that acts like an air-lock entry. The mudroom is small, about 8 ft. by 8 ft., but it's enough to buffer the cold. Many original New England farmhouses featured mudrooms as a buffer between heated indoor air and the cold outdoor air.

The brick masonry woodstove at the center of the house provides most of the heating. A masonry woodstove, sometimes called a Finnish or Russian woodstove, is different from a traditional fireplace or metal woodstove. A masonry woodstove recaptures the heated gases that escape the burning fire and stores this heat in the stove's prodigious thermal mass. Through a series of runs and baffles, smoke goes up, then passes down across the enclosed firebox and goes up again. The stove burns extremely hot, but the brick surface never gets so hot that it burns your hand. Because the brick mass is almost always warm in winter, animals and humans seek it out as a cozy spot during the cold and dreary New England winter.

We elected to build a masonry woodstove because we liked fireplaces but knew they were terribly inefficient. We wanted an efficient heat source but didn't like the way a metal woodstove heats up and cools down quickly, nor did we want to deal with the amount of residue a woodstove creates in the chimney flue. Our masonry woodstove is efficient and clean, and we rarely have to use our backup, oil-fired forced hot water system.

Summer cooling—The house is designed to keep cool through a combination of ventilation and shading. Although the setting winter sun penetrates the house to illuminate the masonry woodstove, the roof's overhangs provide shade during the summer. Building a substantial roof overhang was one way to keep our house cool. Both the soffits and the rakes overhang 2 ft. These large overhangs make it unnecessary to close the upstairs windows during summer storms because rain rarely reaches the upper portion of the house.

The west elevation of the house has a large screen porch (bottom left photo, p. 154) that provides shade where summer overheating potential is greatest. The gable roof on this side of the house changes from a 12-in-12 pitch at the top to a 4-in-12 pitch at the plate where the porch framing begins. The porch roof was originally designed to come down lower, but when we were building the porch, we realized a low-hanging roof would obscure the best part of the view. Raising the ridge height and the eaves wall provided a steep enough pitch on the porch roof and opened up the view.

The west-facing screen porch catches the cool summer breezes coming off the nearby Nashua River. Two sets of French doors connect the living room to the screen porch and allow the cool air from the porch to circulate through the living room.

Along with the operable skylights in the cathedral ceilings, a turret above the second floor expels rising hot air at the high point of the house while drawing cool air from the porch and the garden room, where open windows and shading from deciduous trees control the temperature. The result is convection air movement, where rising hot air is replaced by cool air.

You must climb a ladder from a second-story study to reach the turret, but once there you have a 270° view around you through more single-pane, double-hung windows. A skylight in the hip-roof cathedral ceiling offers great views of the stars at night. The turret even has a pine floor. Although this room measures 8 ft. by 8 ft. (the floor opening is about 2 ft. by 3 ft.), the house sits at the highest point of the lot, so you feel as if you are sitting on top of the world when you're in the turret. □

Linda Moody is an architect in Pepperell, Mass. Photos by Rich Ziegner.

Index

The articles in this book originally appeared in *Fine Homebuilding* magazine. The date of first publication, issue number, and page numbers for each article are given at right.